CRC Handbook of Fundamental Spectroscopic Correlation Charts

THOMAS J. BRUNO

PARIS D. N. SVORONOS

Taylor & Francis

Taylor & Francis Group

Boca Raton   London   New York

A CRC title, part of the Taylor & Francis imprint, a member of the
Taylor & Francis Group, the academic division of T&F Informa plc.

Certain commercial equipment, instruments, or materials are identified in this handbook in order to provide an adequate description. Such identification does not imply recommendation or endorsement by the National Institute of Standards and Technology, the City University of New York, or Georgetown University, nor does it imply that the materials or equipment identified are necessarily the best available for the purpose. The authors, publishers, and their respective institutions are not responsible for the use to which this handbook is made. Occasion use is made of non-SI units to conform to the standard and accepted practice in modern analytical and organic chemistry.

Published in 2006 by
CRC Press
Taylor & Francis Group
6000 Broken Sound Parkway NW, Suite 300
Boca Raton, FL 33487-2742

Not subject to copyright in the United States.
CRC Press is an imprint of Taylor & Francis Group

No claim to original U.S. Government works
Printed in the United States of America on acid-free paper
10 9 8 7 6 5 4 3 2 1

International Standard Book Number-10: 0-8493-3250-8 (Softcover)
International Standard Book Number-13: 978-0-8493-3250-0 (Softcover)
Library of Congress Card Number 2005050553

This book contains information obtained from authentic and highly regarded sources. Reprinted material is quoted with permission, and sources are indicated. A wide variety of references are listed. Reasonable efforts have been made to publish reliable data and information, but the author and the publisher cannot assume responsibility for the validity of all materials or for the consequences of their use.

Library of Congress Cataloging-in-Publication Data

Bruno, Thomas J.
    CRC handbook of fundamental spectroscopic correlation charts / authors, Thomas J. Bruno, Paris D.N. Svoronos.
        p. cm.
    Includes bibliographical references and indexes.
    ISBN 0-8493-3250-8 (alk. paper)
    1. Spectrum analysis--Statistical methods--Handbooks, manuals, etc. 2. Spectrum analysis--Statistical methods--Tables. I. Svoronos, Paris D.N. II. Title.

QD95.5.S72B78 2005
543'.5--dc22   2005050553

Taylor & Francis Group
is the Academic Division of Informa plc.

Visit the Taylor & Francis Web site at
http://www.taylorandfrancis.com

and the CRC Press Web site at
http://www.crcpress.com

Dedication

We dedicate this work to our children, Kelly-Anne, Alexandra, and Theodore.

Preface

This work began as a slim booklet prepared by one of the authors (TJB) to accompany a course on chemical instrumentation presented at the National Institute of Standards and Technology, Boulder Laboratories. The booklet contained tables and charts on chromatography, spectroscopy, and chemical (wet) methods, and was intended to provide the students with enough basic data to design their own analytical methods and procedures. Shortly thereafter, with the coauthorship of Professor Paris D.N. Svoronos, it was expanded into a more extensive compilation entitled "Basic Tables for Chemical Analysis," published as a National Institute of Standards and Technology Technical Note (No. 1096), and then was further expanded to become the *CRC Handbook of Basic Tables for Chemical Analysis*. This highly successful book is now in its second edition.

In both of those editions, spectroscopic correlations are presented essentially in tabular form. It is our conviction, however, that for spectroscopic data, correlation charts are far more easily used than tables. This is especially true of scientists and students who must make use (occasional or frequent) of spectroscopic methods, while their own areas of specialization may lie elsewhere. We have therefore developed a set of fundamental correlation charts that target the same audience as the tabular format adopted in *CRC Handbook of Basic Tables for Chemical Analysis*. The exception would be the tables that we present here for mass spectrometry, a technique that does not lend itself to graphical presentation of data. It is nevertheless important to have data for this important technique covered here, since often multiple techniques, including mass spectrometry, are used for analyses and structural determinations.

Our philosophy in preparing this book has been to include only information that will help the user make decisions. In this respect, we envision each chart as being something the user will consult when reaching a decision point while interpreting spectroscopic results. We have deliberately chosen to exclude information that is merely interesting, but of little value at a decision point.

Similarly, it has occasionally been difficult to strike an appropriate balance between presenting information that is of general utility, and information that is highly specific. We realize that there are multivolume sets of reference works that provide comprehensive, if not exhaustive, spectral charts on specific compound classes. Clearly, this book is not meant to replace these sources. In contrast, this book fills the need for fundamental charts that are needed on a general, day-to-day basis. In this respect, we have tried to keep the content as generic as possible. This must not be regarded as a value judgment, but simply as a reflection of our philosophy. The format of the book, a spiral-bound volume that will lie flat on the lab bench, is very much in keeping with our purposes. We are hoping for a volume that will "go to work" each day with the user.

Thomas J. Bruno
Paris D.N. Svoronos

Acknowledgments

The authors would like to acknowledge some individuals who have been of great help during the preparation of this work. Marilyn Yetzbacher of NIST prepared much of the artwork used throughout this volume. Without her help and expertise, this volume could never have been completed. In addition, Stephanie Outcalt, a chemical engineer at NIST, prepared some of the original correlation charts used here and in previous volumes. We owe a great debt to our board of reviewers: Drs. John Coates, J. Widegren, D. Joshi, S. Ghayourmanesh, D.G. Friend, M. Huber, J. Marino, S. Ringen, N. Sari, D. Smith, and B. Cage and Profs. A.F. Lagalante, K.E. Miller, and K. Nakanishi. Finally, we must again thank our wives, Clare and Soraya, and our children, Kelly-Anne, Alexandra, and Theodore, for their patience and support throughout the period of hard work and late nights.

The Authors

Thomas J. Bruno, Ph.D., is a project leader in the Physical and Chemical Properties Division at the National Institute of Standards and Technology, Boulder, Colorado. Dr. Bruno received his B.S. in chemistry from the Polytechnic Institute of Brooklyn, and his M.S. and Ph.D. in physical chemistry from Georgetown University. He served as a National Academy of Sciences–National Research Council postdoctoral associate at NIST, and was later appointed to the staff. Dr. Bruno has done research on properties of fuel mixtures, chemically reacting fluids, and environmental pollutants. He is also involved in research on supercritical fluid extraction and chromatography of bioproducts, the development of novel analytical methods for environmental contaminants and alternative refrigerants, novel detection devices for chromatography, and he manages the division's analytical chemistry laboratory. In his research areas, he has published approximately 125 papers and 7 books, and he holds 10 patents. He was awarded the Department of Commerce Bronze Medal in 1986 for his work on the thermophysics of reacting fluids. He has served as a forensic consultant and an expert witness for the U.S. Department of Justice (DOJ), and received a letter of commendation from DOJ for these efforts in 2002.

Paris D.N. Svoronos, Ph.D., is professor of chemistry and department chair at QCC of the City University of New York. In addition, he holds a continuing appointment as visiting professor in the Department of Chemistry at Georgetown University. Dr. Svoronos obtained a B.S. in chemistry and a B.S. in physics at the American University of Cairo, and his M.S. and Ph.D. in organic chemistry at Georgetown University. Among his research interests are synthetic sulfur and natural product chemistry, organic electrochemistry, and organic structure determination and trace analysis. He also maintains a keen interest in chemical education, and has authored several laboratory manuals used widely at the undergraduate level. In his fields of interest, he has approximately 70 publications. He has been included in the *Who's Who of America's Teachers* five times in the last six years. He is particularly proud of his students' successes in research presentations, paper publications, and professional accomplishments. He was selected as the 2003 Outstanding Professor of the Year by CASE (Council for the Advancement and Support of Education) and the Carnegie Foundation.

Contents

CHAPTER 1

Ultraviolet–Visible Spectrophotometry

CONTENTS

CORRELATION CHARTS FOR ULTRAVIOLET–VISIBLE SPECTROPHOTOMETRY

This section provides correlation charts and operational information for the design and interpretation of ultraviolet–visible spectrophotometric (UV-Vis) measurements.[1–4] While UV-Vis is perhaps not as information-rich as infrared or nuclear magnetic resonance, it nonetheless has value in structure determination and sample identification. Moreover, it is extremely valuable in quantitative work. Typical UV-Vis instruments cover not only the UV and visible spectrum, but the near-infrared as well. Although there is overlap among the ranges, the approximate breakdown is:

Ultraviolet: 10–380 nm
    far-ultraviolet 10–200 nm
    near-ultraviolet 185–380 nm
Visible region:
    violet 380–450 nm
    blue 450–495 nm
    green 495–570 nm
    yellow 570–590 nm
    orange 590–620 nm
    red 620–750 nm
Near-infrared: 750–3000 nm

In the older literature, the far-ultraviolet is often referred to as the vacuum ultraviolet. Since oxygen absorbs strongly below 200 nm, the way spectra were originally recorded below this wavelength was to evacuate the spectrophotometer. An alternative has been to purge the instrument with nitrogen, which allows measurement down to 150 nm, at which point nitrogen begins to absorb strongly. The division between the visible and ultraviolet regions results from the light that a typical human can see; the essential spectral processes (electronic transitions) are the same in both regions, however. Information on near-infrared spectrometry is provided in the section on infrared spectrometry. In the correlation charts that follow, the user should keep in mind that the indicated bands are for peak maxima (the apex). Typical UV-Vis spectral bands are broad and cover a wide wavelength range. The charts provide guidance on the location of the peak apex. The extinction coefficient (ε) ranges are also approximate.

REFERENCES

1. Bruno, T.J. and Svoronos, P.D.N., *CRC Handbook of Basic Tables for Chemical Analysis,* 2nd ed., CRC Press, Boca Raton, FL, 2003.
2. Willard, H.H., Merritt, L.L., Dean, J.A., and Settle, F.A., *Instrumental Methods of Analysis,* 7th ed., Wadsworth Publishing Co., Belmont, CA, 1988.
3. Silverstein, R.M. and Webster, F.X., *Spectrometric Identification of Organic Compounds,* 6th ed., Wiley, New York, 1998.
4. Lambert, J.B., Shurvell, H.F., Lightner, D.A., Verbit, L., and Cooks, R.G., *Organic Structural Spectroscopy,* Prentice Hall, Upper Saddle River, NJ, 1998.

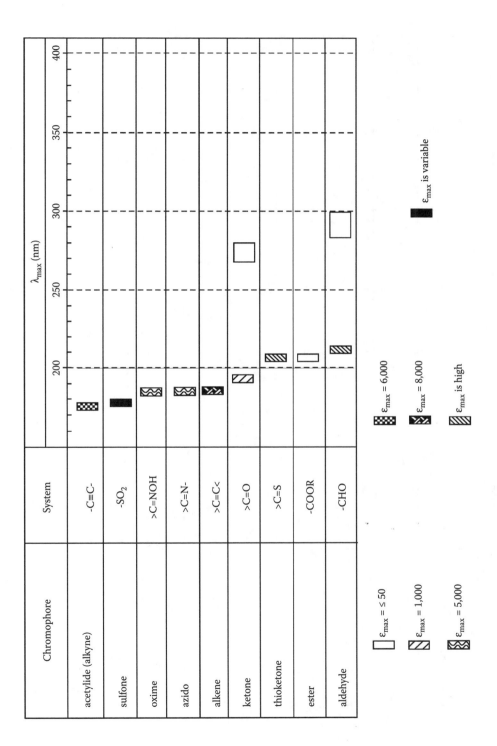

Figure 1.1

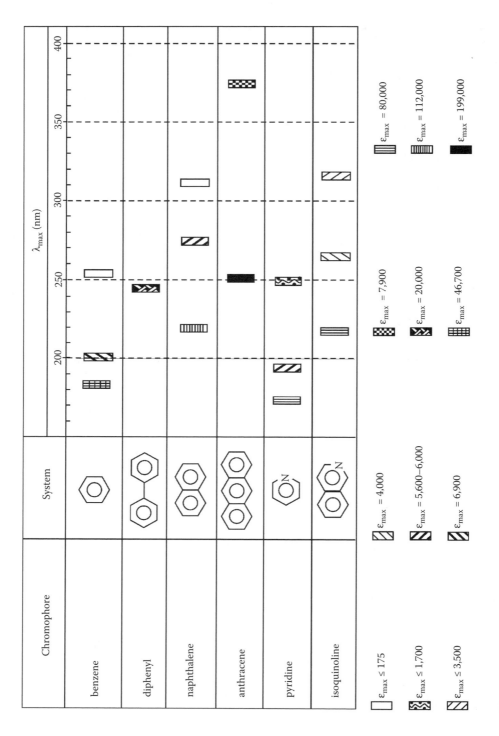

Figure 1.2

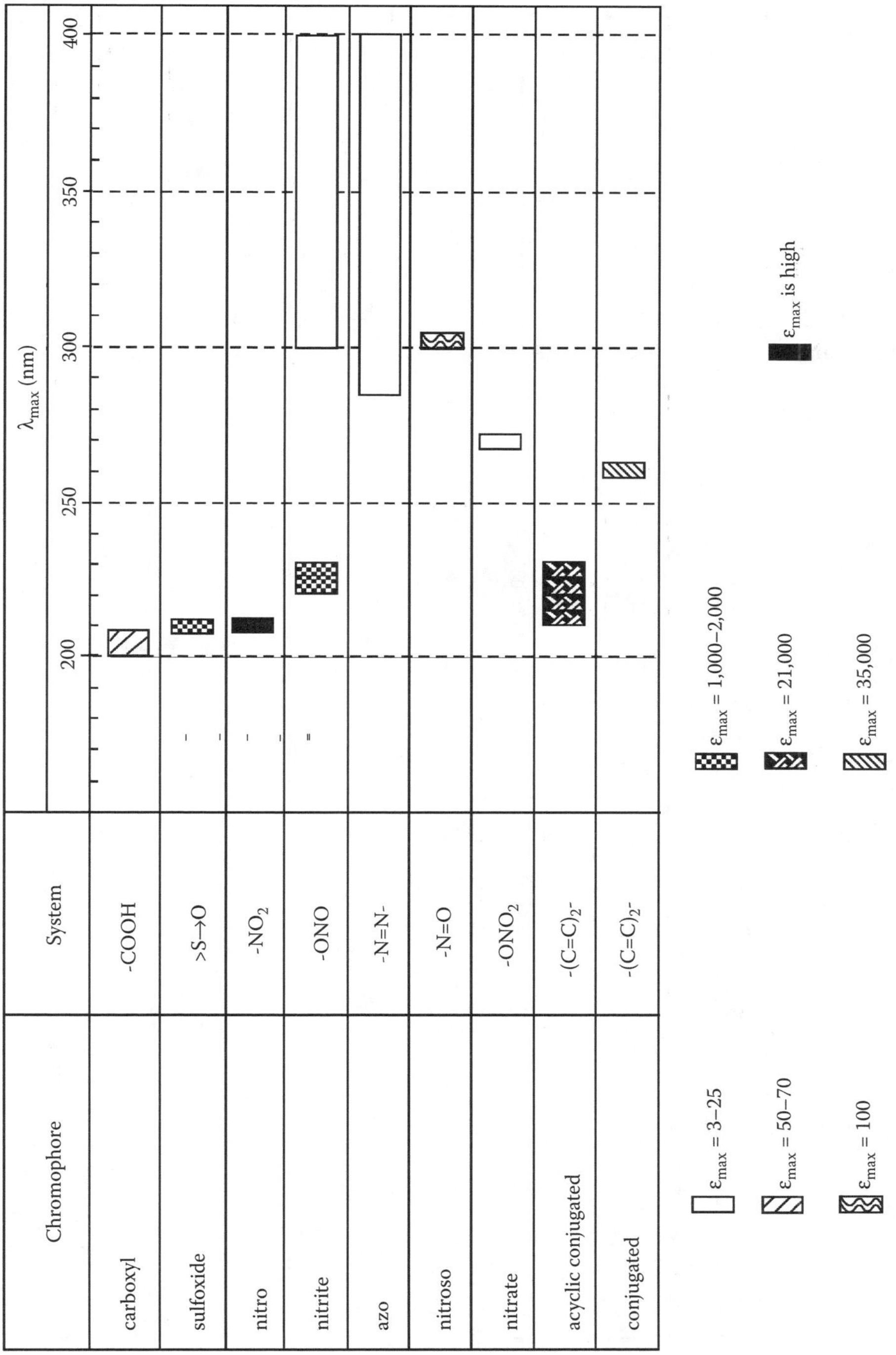

Chromophore
System
λ_{max} (nm)
200
250
300
350
400
carboxyl
-COOH
sulfoxide
>S→O
nitro
$-NO_2$
nitrite
-ONO
azo
-N=N-
nitroso
-N=O
nitrate
$-ONO_2$
acyclic conjugated
$-(C=C)_2-$
conjugated
$-(C=C)_2-$
ε_{max} = 3–25
ε_{max} = 50–70
ε_{max} = 100
ε_{max} = 1,000–2,000
ε_{max} = 21,000
ε_{max} = 35,000
ε_{max} is high

Figure 1.3

Figure 1.4

Chromophore	System	λ_{max} (nm)
conjugated	$-(C=C)_4-$	
conjugated	$-(C=C)_5-$	
alicyclic conjugated	$-(C=C)_2-$	
conjugated	$>C=C-C\equiv C-$	
conjugated	$>C=C-C=N-$	
conjugated	$>C=C-C=O$	
conjugated	$>C=C-NO_2$	
quinoline		

Legend:
- ε_{max} is weak
- $\varepsilon_{max} = 2,750$
- $\varepsilon_{max} = 3,600$
- $\varepsilon_{max} = 3,000–8,000$
- $\varepsilon_{max} = 6,500$
- $\varepsilon_{max} = 9,500$
- $\varepsilon_{max} = 10,000–20,000$
- $\varepsilon_{max} = 23,000$
- $\varepsilon_{max} = 37,000$
- $\varepsilon_{max} = 52,000$
- $\varepsilon_{max} = 118,000$

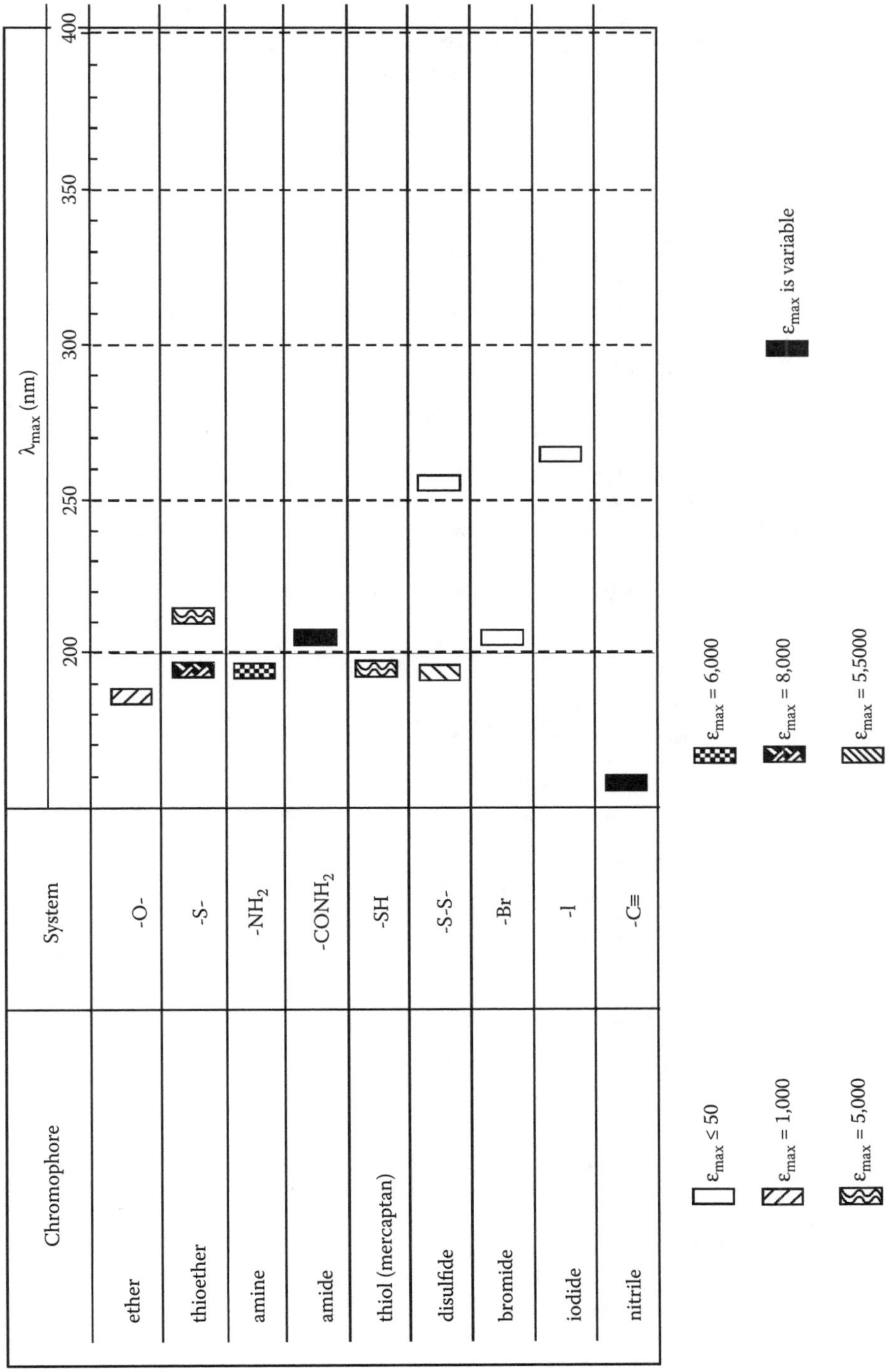

Chromophore
System
λ_{max} (nm)
200
250
300
350
400
ether
-O-
thioether
-S-
amine
$-NH_2$
amide
$-CONH_2$
thiol (mercaptan)
-SH
disulfide
-S-S-
bromide
-Br
iodide
-I
nitrile
-C≡
$\varepsilon_{max} \leq 50$
ε_{max} = 1,000
ε_{max} = 5,000
ε_{max} = 6,000
ε_{max} = 8,000
ε_{max} = 5,5000
ε_{max} is variable

Figure 1.5

WOODWARD'S RULES FOR BATHOCHROMIC SHIFTS

Conjugated systems show bathochromic shifts in their $\pi \rightarrow \pi^*$ transition bands. Empirical methods for predicting those shifts were originally formulated by Woodward,[1-3] and Fieser and Fieser.[4] This section includes the most important conjugated system rules.[1-7] The increments here are given in nanometers (nm). The reader should consult references 5 and 6 for more details on how to apply the wavelength increment data.

REFERENCES

1. Woodward, R.B., Structure and the absorption spectra of α,β-unsaturated ketones, *J. Am. Chem. Soc.,* 63, 1123, 1941.
2. Woodward, R.B., Structure and absorption spectra, III: Normal conjugated dienes, *J. Am. Chem. Soc.,* 64, 72, 1942.
3. Woodward, R.B., Structure and absorption spectra, IV: Further observations on α,β-unsaturated ketones, *J. Am. Chem. Soc.,* 64, 76, 1942.
4. Fieser, L.F. and Fieser, M., *Natural Products Related to Phenanthrene,* Reinhold, New York, 1949.
5. Silverstein, R.M. and Webster, F.X., *Spectrometric Identification of Organic Compounds,* 6th ed., John Wiley and Sons, New York, 1998.
6. Lambert, J.B., Shurvell, H.F., Lightner, D.A., Verbit, L., and Cooks, R.G., *Organic Structural Spectroscopy,* Prentice Hall, Upper Saddle River, NJ, 1998.
7. Bruno, T.J and Svoronos, P.D.N., *CRC Handbook of Basic Tables for Chemical Analysis,* 2nd ed., CRC Press, Boca Raton, FL, 2003.

Rules for Diene Absorption

Base value for diene, nm	214
Increments for each, nm	
Heteroannular diene	+0
Homoannular diene	+39
Extra double bond	+30
Alkyl substituent or ring residue	+5
Exocyclic double bond	+5
Polar groups	
−OOCR	+0
−OR	+6
−S−R	+30
Halogen	+5
−NR$_2$	+60
λ calculated =	Total

Rules for Enone Absorption

$$\overset{\delta}{-}C = \overset{\gamma}{C} - \overset{\beta}{C} = \overset{\alpha}{C} - \underset{\underset{O}{\|}}{C} -$$

Base values for enone, nm:	
Acyclic (or six-membered) α,β-unsaturated ketone	215
Five-membered α,β-unsaturated ketone	202
α,β-unsaturated aldehydes	210
α,β-unsaturated esters or carboxylic acids	195

(continued)

Rules for Enone Absorption (continued)

Increments for each, nm	
Heteroannular diene	+0
Homoannular diene	+39
Double bond	+30
Alkyl group	
$\alpha-$	+10
$\beta-$	+12
$\gamma-$ and higher	+18
Polar groups	
–OH	
$\alpha-$	+35
$\beta-$	+30
$\delta-$	+50
–OOCR	
$\alpha,\beta,\gamma,\delta$	+6
–OR	
$\alpha-$	+35
$\beta-$	+30
$\gamma-$	+17
$\delta-$	+31
–SR	
$\beta-$	+85
–Cl	
$\alpha-$	+15
$\beta-$	+12
–Br	
$\alpha-$	+25
$\beta-$	+30
$-NR_2$	
$\beta-$	+95
Exocyclic double bond	+5
λ calculated =	Total

Note: Solvent corrections should be included. These are: water (–8), chloroform (+1), dioxane (+5), ether (+7), hexane (+11), cyclohexane (+11). No correction for methanol or ethanol.

Rules for Monosubstituted Benzene Derivatives

Base value for parent chromophore (benzene), nm	250

Substituent	Increment, nm
–R	–4
–COR	–4
–CHO	0
–OH	–16
–OR	–16
–COOR	–16

Note: R indicates an alkyl group, and the substitution is on the benzene ring, $C_6H_5–$.

Rules for Disubstituted Benzene Derivatives

Base value for parent chromophore (benzene), nm	250

Substituent	Increment, nm		
	o-	m-	p-
–R	+3	+3	+10
–COR	+3	+3	+10
–OH	+7	+7	+25
–OR	+7	+7	+25
–O⁻	+11	+20	+78 (variable)
–Cl	+0	+0	+10
–Br	+2	+2	+15
–NH$_2$	+13	+13	+58
–NHCOCH$_3$	+20	+20	+45
–NHCH$_3$	—	—	+73
–N(CH$_3$)$_2$	+20	+20	+85

Note: R indicates an alkyl group.

SOLVENTS FOR ULTRAVIOLET SPECTROPHOTOMETRY

The following table lists some useful solvents for ultraviolet spectrophotometry, along with their wavelength cutoffs and dielectric constants.[1–7]

REFERENCES

1. Willard, H.H., Merritt, L.L., Dean, J.A., and Settle, F.A., *Instrumental Methods of Analysis,* 7th ed., Van Nostrand, New York, 1988.
2. Strobel, H.A. and Heinemann, W.R., *Chemical Instrumentation: A Systematic Approach,* 3rd ed., John Wiley and Sons, New York, 1989.
3. Dreisbach, R.R., *Physical Properties of Chemical Compounds,* in Advances in Chemistry Series, No. 15, American Chemical Society, Washington, DC, 1955.
4. Dreisbach, R.R., *Physical Properties of Chemical Compounds,* in Advances in Chemistry Series, No. 22, American Chemical Society, Washington, DC, 1959.
5. Sommer, L., *Analytical Absorption Spectrophotometry in the Visible and Ultraviolet,* Elsevier Science, New York, 1989.
6. Krieger, P.A., *High Purity Solvent Guide,* Burdick & Jackson, McGaw Park, IL, 1984.
7. Bruno, T.J. and Svoronos, P.D.N., *CRC Handbook of Basic Tables for Chemical Analysis,* 2nd ed., CRC Press, Boca Raton, FL, 2003.

Solvents for Ultraviolet Spectrophotometry

Solvent	Wavelength Cutoff, nm	Dielectric Constant (20°C)[a]
Acetic acid	260	6.15
Acetone	330	20.7 (25°C)
Acetonitrile	190	37.5
Benzene	280	2.284
sec-Butyl alcohol (2-butanol)	260	15.8 (25°C)
n-Butyl acetate	254	5.07
n-Butyl chloride	220	7.39 (25°C)
Carbon disulfide	380	2.641
Carbon tetrachloride	265	2.238
Chloroform[b]	245	4.806
Cyclohexane	210	2.023
1,2-Dichloroethane	226	10.19 (25°C)
1,2-Dimethoxyethane	240	7.2
N,N-Dimethylacetamide	268	59 (83°C)
N,N-Dimethylformamide	270	36.7
Dimethylsulfoxide	265	4.7
1,4-Dioxane	215	2.209 (25°C)
Diethyl ether	218	4.335
Ethanol	210	24.30 (25°C)
2-Ethoxyethanol	210	—
Ethyl acetate	225	6.02 (25°C)
Methyl ethyl ketone	330	18.5
Glycerol	207	42.5 (25°C)
n-Hexadecane	200	2.06 (25°C)
n-Hexane	210	1.890
Methanol	210	32.63 (25°C)
2-Methoxyethanol	210	16.9
Methyl cyclohexane	210	2.02 (25°C)
Methyl isobutyl ketone	335	13.11 (25°C)

(*continued*)

Solvents for Ultraviolet Spectrophotometry (Continued)

Solvent	Wavelength Cutoff, nm	Dielectric Constant (20°C)[a]
2-Methyl-1-propanol	230	17.7
N-Methyl-2-pyrrolidone	285	32.0
n-Pentane	210	1.844
n-Pentyl acetate	212	5.0
n-Propyl alcohol	210	20.1 (25°C)
sec-Propyl alcohol	210	18.3 (25°C)
Pyridine	330	12.3 (25°C)
Tetrachloroethylene[c]	290	5.1 (19°C)
Tetrahydrofuran	220	7.6
Toluene	286	2.379 (25°C)
1,1,2-Trichloro-1,2,2-trifluoroethane	231	—
2,2,4-Trimethylpentane	215	1.936 (25°C)
o-Xylene	290	2.568
m-Xylene	290	2.374
p-Xylene	290	2.270
Water	—	78.54 (25°C)

[a] Unless otherwise indicated.
[b] Stabilized with ethanol to avoid phosgene formation.
[c] Stabilized with thymol (isopropyl *m*-cresol).

TRANSMITTANCE–ABSORBANCE CONVERSION

The following is a conversion table for absorbance and transmittance, assuming no reflection. Included for each pair is the percent uncertainty propagated into a measured concentration (using the Beer–Lambert Law), assuming an uncertainty in transmittance of -0.005.[1] The value of transmittance that will give the lowest percent uncertainty in concentration is 0.368, with a corresponding absorbance of 0.434. Where possible, analyses should be designed for the low-uncertainty area.

REFERENCE

1. Kennedy, J.H., *Analytical Chemistry Principles,* Harcourt, Brace and Jovanovich, San Diego, 1984.

Conversion Table for Transmittance and Absorbance

Transmittance	Absorbance	Percent Uncertainty
0.980	0.009	25.242
0.970	0.013	16.915
0.960	0.018	12.752
0.950	0.022	10.256
0.940	0.027	8.592
0.930	0.032	7.405
0.920	0.036	6.515
0.910	0.041	5.823
0.900	0.046	5.270
0.890	0.051	4.818
0.880	0.056	4.442
0.870	0.060	4.125
0.860	0.065	3.853
0.850	0.071	3.618
0.840	0.076	3.412
0.830	0.081	3.231
0.820	0.086	3.071
0.810	0.091	2.928
0.800	0.097	2.799
0.790	0.102	2.684
0.780	0.108	2.579
0.770	0.113	2.483
0.760	0.119	2.386
0.750	0.125	2.316
0.740	0.131	2.243
0.730	0.137	2.175
0.720	0.143	2.113
0.710	0.149	2.055
0.700	0.155	2.002
0.690	0.161	1.952
0.680	0.167	1.906
0.670	0.174	1.863
0.660	0.180	1.822
0.650	0.187	1.785
0.640	0.194	1.750
0.630	0.201	1.717
0.620	0.208	1.686
0.610	0.215	1.657
0.600	0.222	1.631
0.590	0.229	1.605
0.580	0.237	1.582

(continued)

Conversion Table for Transmittance and Absorbance (Continued)

Transmittance	Absorbance	Percent Uncertainty
0.570	0.244	1.560
0.560	0.252	1.539
0.540	0.268	1.502
0.530	0.276	1.485
0.520	0.284	1.470
0.510	0.292	1.455
0.500	0.301	1.442
0.490	0.310	1.430
0.480	0.319	1.419
0.470	0.328	1.408
0.460	0.337	1.399
0.450	0.347	1.391
0.440	0.356	1.383
0.430	0.366	1.377
0.420	0.377	1.372
0.410	0.387	1.367
0.400	0.398	1.364
0.390	0.409	1.361
0.380	0.420	1.359
0.370	0.432	1.358
0.360	0.444	1.359
0.350	0.456	1.360
0.340	0.468	1.362
0.330	0.481	1.366
0.320	0.495	1.371
0.310	0.509	1.376
0.300	0.523	1.384
0.290	0.538	1.392
0.280	0.553	1.402
0.270	0.569	1.414
0.260	0.585	1.427
0.250	0.602	1.442
0.240	0.620	1.459
0.230	0.638	1.478
0.220	0.657	1.500
0.210	0.678	1.525
0.200	0.699	1.553
0.190	0.721	1.584
0.180	0.745	1.619
0.170	0.769	1.659
0.160	0.796	1.704
0.150	0.824	1.756
0.140	0.854	1.816
0.130	0.886	1.884
0.120	0.921	1.964
0.110	0.958	2.058
0.100	1.000	2.170
0.090	1.046	2.306
0.080	1.097	2.473
0.070	1.155	2.685
0.060	1.222	2.961
0.050	1.301	3.336
0.040	1.398	3.881
0.030	1.523	4.751
0.020	1.699	6.387
0.010	2.000	10.852

Note: Assuming no reflection.

CALIBRATION OF ULTRAVIOLET–VISIBLE SPECTROPHOTOMETERS

The following is not strictly a "calibration table." More accurately, this table provides guidance in checking the wavelength reliability of typical ultraviolet–visible (UV-Vis) spectrophotometers.[1] Calibration is usually not straightforward and is typically done in the facilities of spectrophotometer manufacturers.

There are three primary methods to check the wavelength uncertainty of UV-Vis instruments. The first involves the placement of a standard glass filter (usually provided by or available from the instrument manufacturer) in the sample beam. These filters are common rare earth (holmium and didymium, a mixed glass of neodymium and praseodymium) oxide glasses that produce a series of sharp absorbance bands. This method is useful despite the variation that can sometimes occur between batches of glasses. Variation is reflected in the uncertainty statement provided with each wavelength. The following table provides some of the major absorbances of these glasses.

Wavelengths of Absorbances in the UV-Vis of Rare Earth Glasses, nm

Holmium Glass	Didymium Glass
241.5 ± 0.2	573.0 ± 3.0
279.4 ± 0.3	586.3 ± 3.0
287.5 ± 0.4	685.0 ± 4.5
333.7 ± 0.6	—
360.9 ± 0.8	—
418.4 ± 1.1	—
453.2 ± 1.4	—
536.2 ± 2.3	—
637.5 ± 3.8	—

A more reliable method to check the wavelength uncertainty in UV-Vis instruments is to insert a mercury- or neon-discharge lamp into the sample compartment and then to measure the wavelengths of the sharp lines that result. The following table has the lines that are commonly used for reference. These are provided with no uncertainty statement, since the values are typically less uncertain than the repeatability of UV-Vis instruments.

Wavelengths of Neon and Mercury Discharge Lines, nm

Neon Lines	Mercury Lines
533.1	253.7
534.1	364.9
540.1	404.5
585.3	435.8
614.3	546.1
633.4	576.9
640.2	579.0
667.8	—
693.0	—
717.4	—
724.5	—

In addition, a quick check of the wavelength can be done with the deuterium line, since many instruments operate with a deuterium lamp. Two lines are used most commonly:

Deuterium red line: 656.1 nm
Deuterium blue-green line: 486.0 nm

A still more approximate check can be done in the visible region with a small inspection mirror inserted in the sample compartment to observe the light. With a visible (tungsten) source energized, one observes when the light appears to be pure yellow, with no red or green component. The wavelength at this point should be 570 to 580 nm.

REFERENCES

1   Denney, R.C. and Sinclair, R., *Visible and Ultraviolet Spectroscopy: Analytical Chemistry by Open Learning*, John Wiley & Sons, Chichester, 1987.

CHAPTER 2

Infrared Spectrophotometry

CONTENTS

INFRARED ABSORPTION CORRELATION CHARTS

The following charts provide characteristic infrared absorptions obtained from particular functional groups on molecules.[1] These include a general mid-range correlation chart, a chart for aromatic absorptions, and a chart for carbonyl moieties. The general mid-range chart is an adaptation of the work of Prof. Charles F. Hammer of Georgetown University, reproduced with modification and with permission.

REFERENCE

1. Bruno, T.J. and Svoronos, P.D.N., *CRC Handbook of Basic Tables for Chemical Analysis*, 2nd ed., CRC Press, Boca Raton, FL, 2003.

LEGEND FOR CORRELATION CHARTS

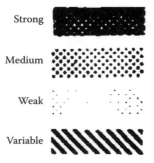

Strong

Medium

Weak

Variable

Note: AR = aromatic, b = broad, sd = solid, sn = solution, sp = sharp, ? = unreliable.

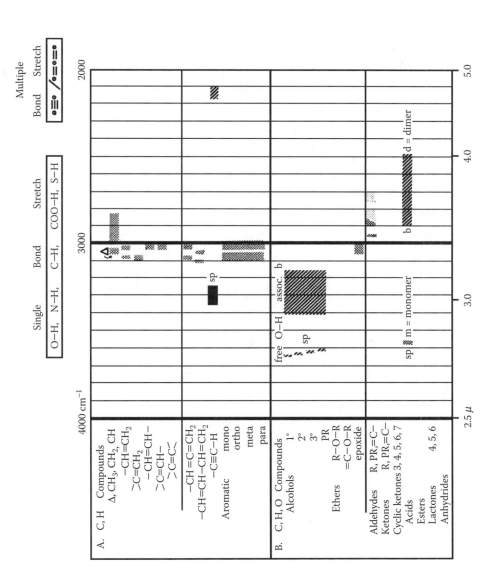

Figure 2.1

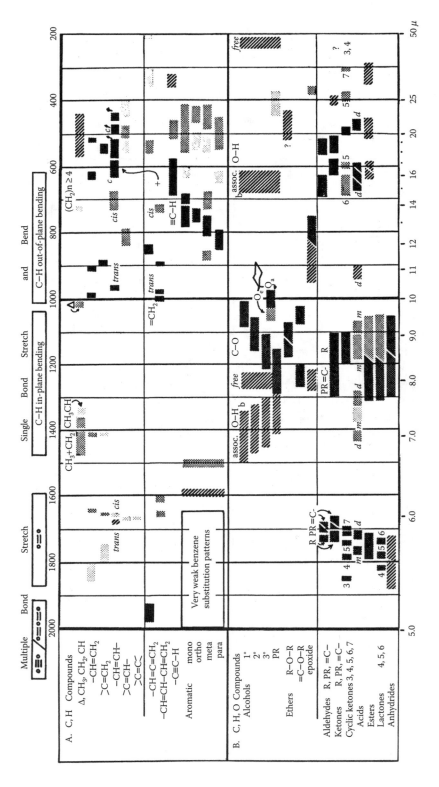

Figure 2.2

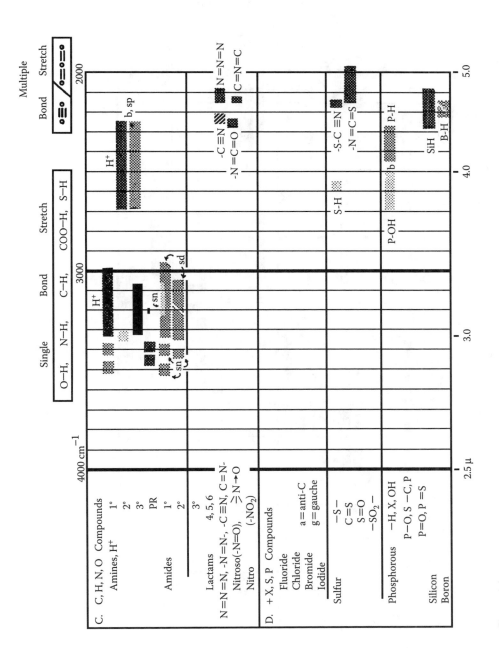

Figure 2.3

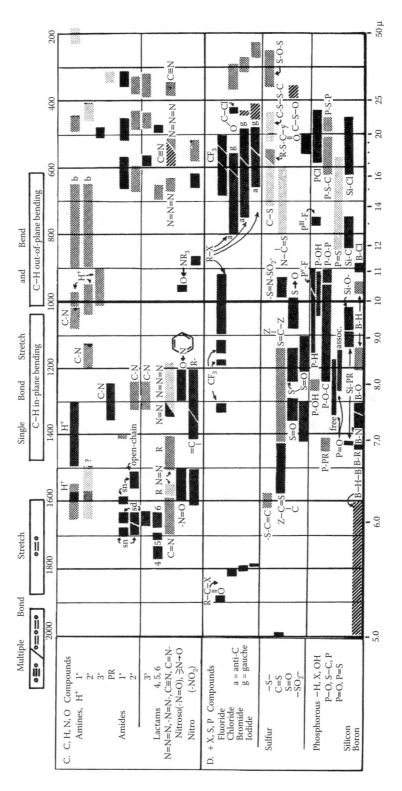

Figure 2.4

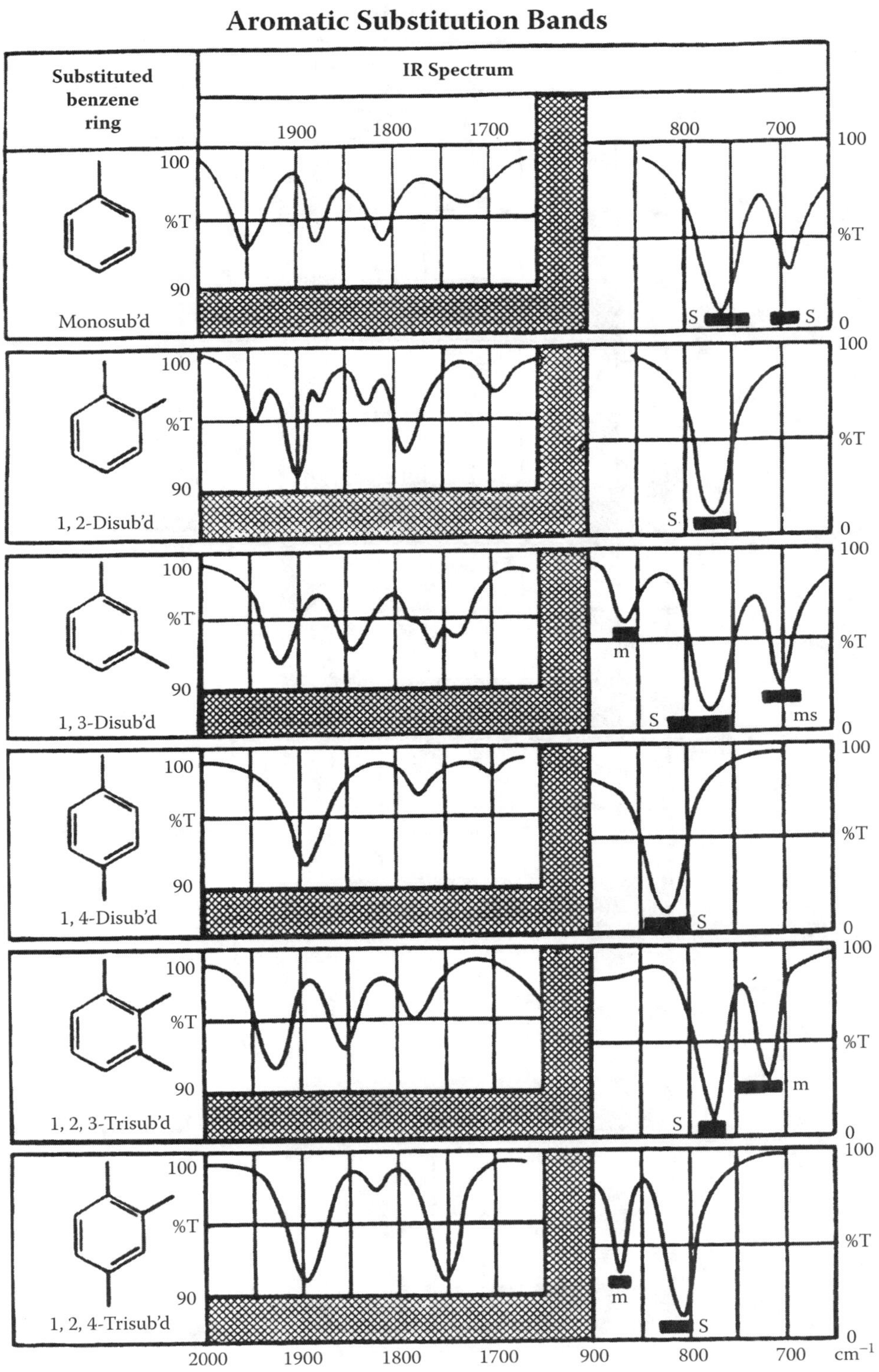

Aromatic Substitution Bands
Substituted benzene ring
IR Spectrum
1900
1800
1700
800
700
100
%T
90
0
Monosub'd
1, 2-Disub'd
1, 3-Disub'd
1, 4-Disub'd
1, 2, 3-Trisub'd
1, 2, 4-Trisub'd
S
m
ms
2000
900
cm^{-1}

Figure 2.5

Aromatic Substitution Bands

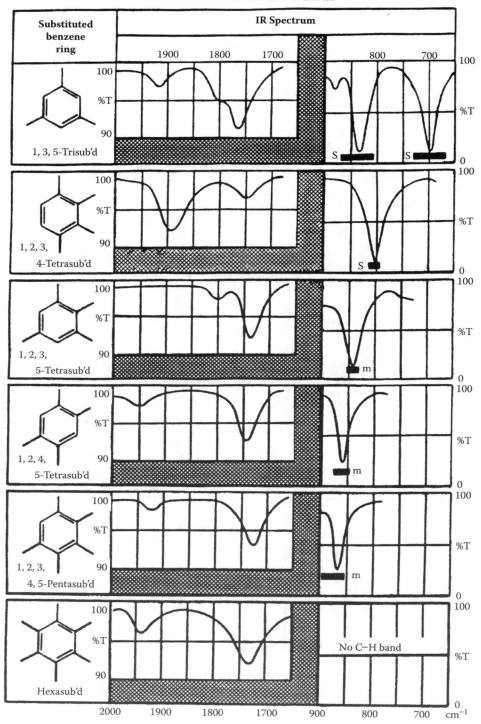

Figure 2.6

Carbonyl Group Absorptions

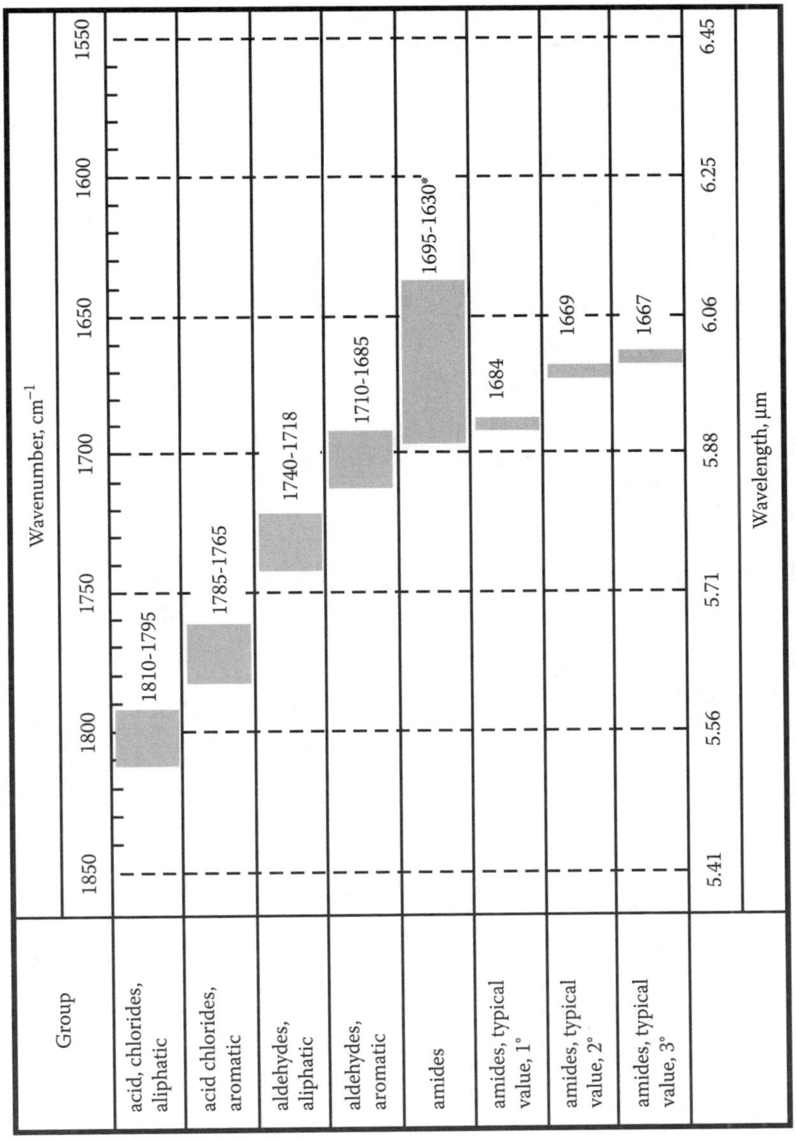

Group	1850	1800	1750	1700	1650	1600	1550
acid, chlorides, aliphatic		1810-1795					
acid chlorides, aromatic			1785-1765				
aldehydes, aliphatic				1740-1718			
aldehydes, aromatic				1710-1685			
amides					1695-1630*		
amides, typical value, 1°				1684			
amides, typical value, 2°					1669		
amides, typical value, 3°					1667		
	5.41	5.56	5.71	5.88	6.06	6.25	6.45

Wavenumber, cm^{-1}

Wavelength, μm

* Electron withdrawing groups at the α-position to the carbonyl will raise the wavenumber of the absorption.

Figure 2.7

Carbonyl Group Absorptions (continued)

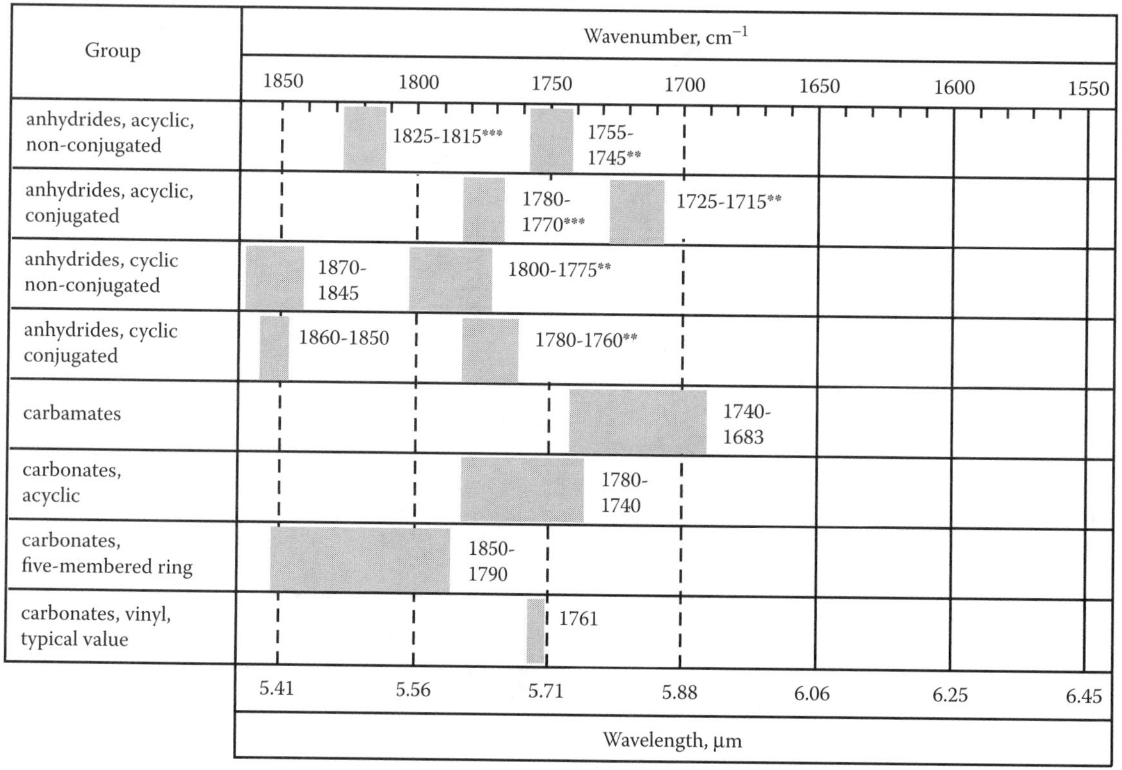

** This band is the more intense of the two.
*** Intensity weakens as colinearity is approached.

Figure 2.8

Carbonyl Group Absorptions (continued)

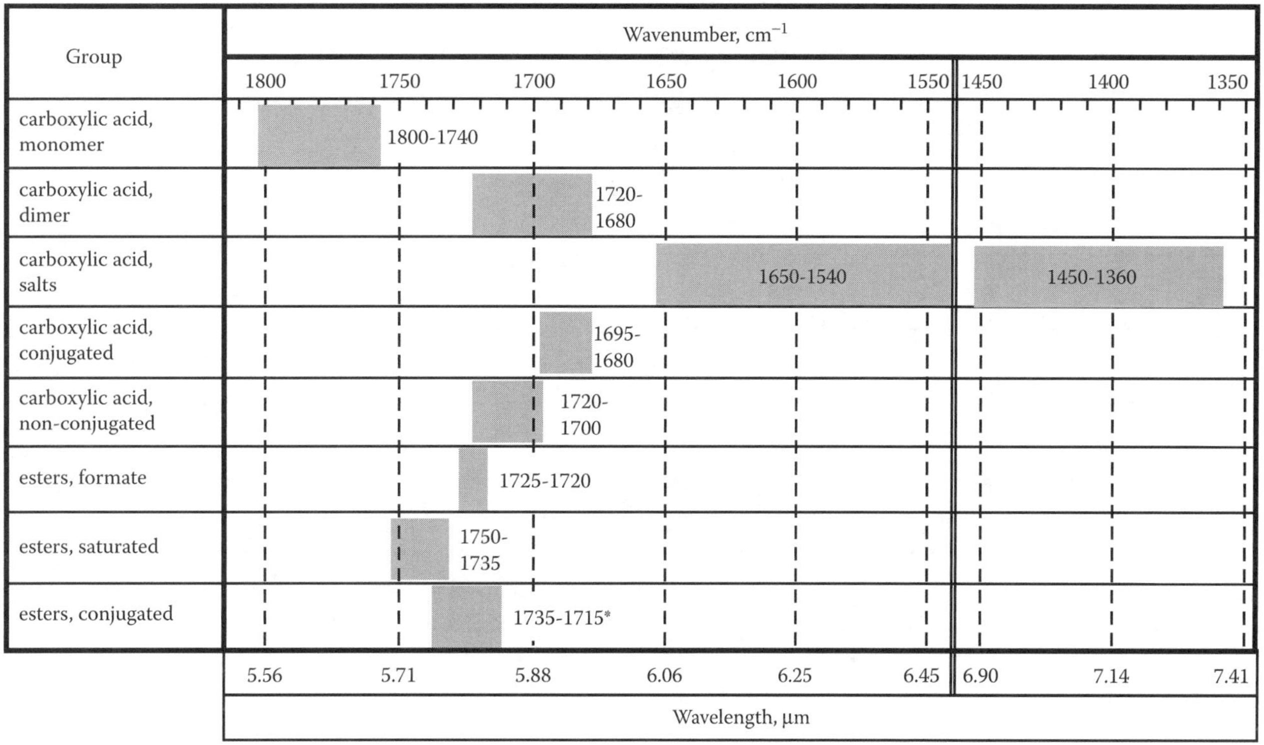

Group	Wavenumber, cm⁻¹									
	1800	1750	1700	1650	1600	1550	1450	1400	1350
carboxylic acid, monomer	1800-1740								
carboxylic acid, dimer			1720-1680						
carboxylic acid, salts					1650-1540		1450-1360		
carboxylic acid, conjugated			1695-1680						
carboxylic acid, non-conjugated			1720-1700						
esters, formate			1725-1720						
esters, saturated		1750-1735							
esters, conjugated		1735-1715*							
	5.56	5.71	5.88	6.06	6.25	6.45	6.90	7.14	7.41

Wavelength, μm

* Electron withdrawing groups in the α-position to the carbonyl will raise the wavenumber adsorption.

Figure 2.9

Carbonyl Group Absorptions (continued)

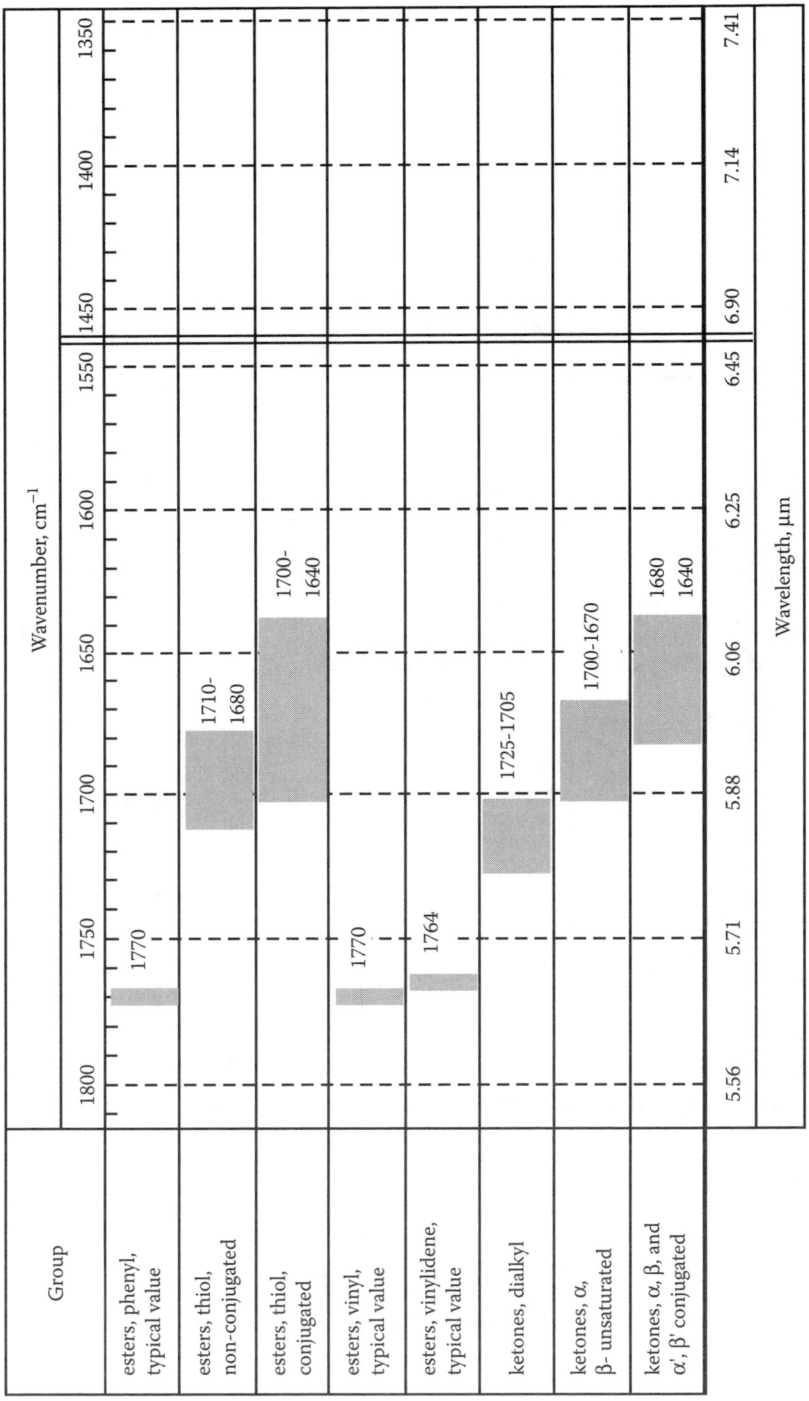

Figure 2.10

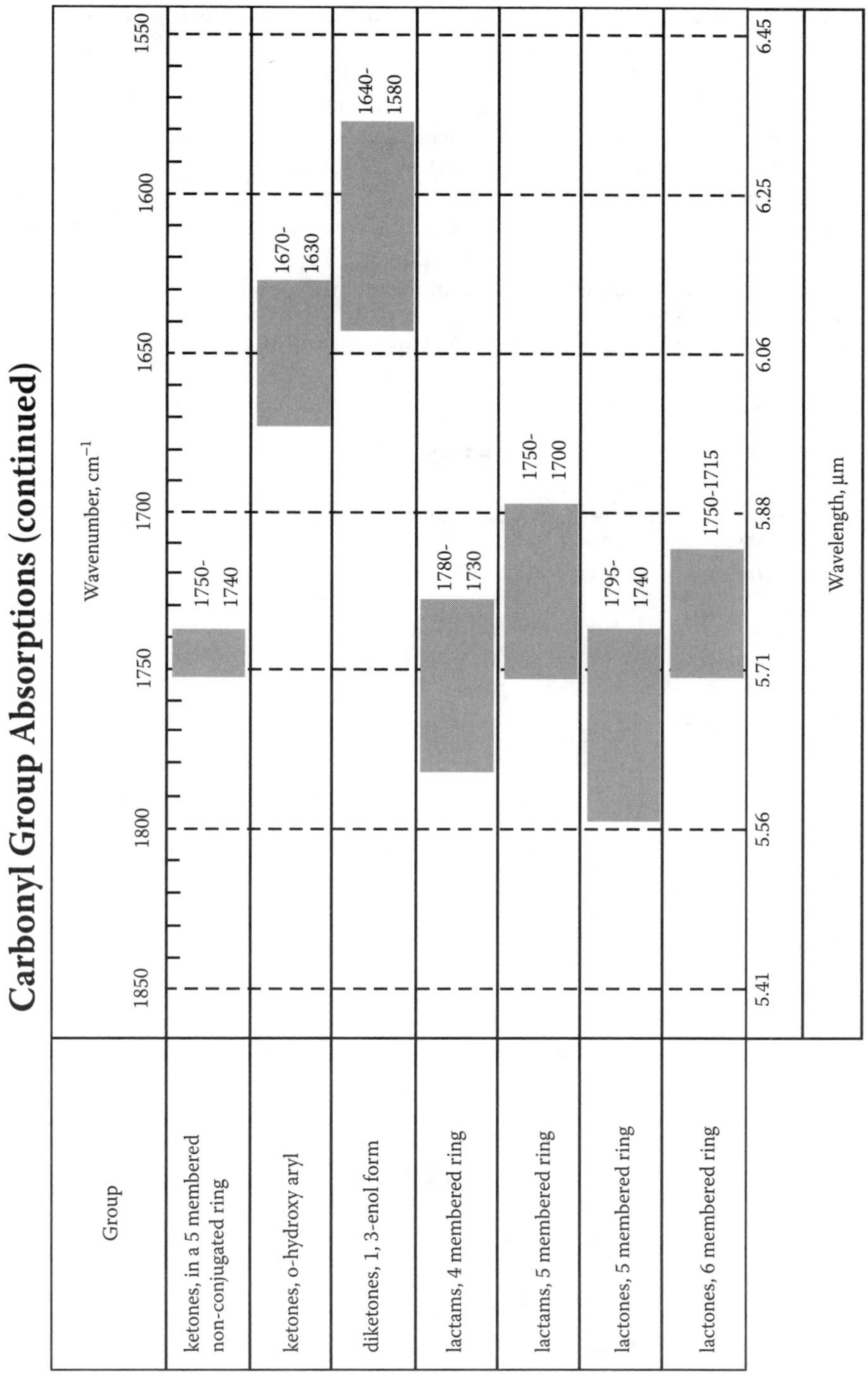
Carbonyl Group Absorptions (continued)
Group
Wavenumber, cm^{-1}
1850
1800
1750
1700
1650
1600
1550
ketones, in a 5 membered non-conjugated ring
1750-1740
ketones, o-hydroxy aryl
1670-1630
diketones, 1, 3-enol form
1640-1580
lactams, 4 membered ring
1780-1730
lactams, 5 membered ring
1750-1700
lactones, 5 membered ring
1795-1740
lactones, 6 membered ring
1750-1715
5.41
5.56
5.71
5.88
6.06
6.25
6.45
Wavelength, μm

Figure 2.11

NEAR-INFRARED ABSORPTIONS

Classically, the near-infrared (NIR) region was defined as occurring between 0.7 and 3.5 μm, or 14,285 to 2,860 cm^{-1}. This classification includes the region of CH, OH, and NH fundamental stretching bands.[1,2] Currently, this spectral area, from 4000 to 2860 cm^{-1}, is considered part of the mid-infrared region, and the NIR region is now considered to be above 4000 cm^{-1}. The NIR is a region of overtones and combination bands, which are considerably weaker than the fundamentals that are seen in the mid-infrared region. It is nevertheless a very useful area for quantitative measurement, on-line and at-line analysis, the analysis of viscous liquids and powders, and even for structure determination.

Because most NIR spectrophotometers are often built as enhancements to the capabilities of ultraviolet–visible spectrophotometers, the convention has been to express absorbances in this region in terms of wavelength rather than wavenumber. In the following charts, we adopt this convention. We do not give any indication of intensity in these charts. The NIR bands will be related to the intensity of the fundamentals in the mid-infrared region, although the bands will typically be broad.

REFERENCES

1. Colthup, N.B., Daly, L.H., and Wiberley, S.E., *Introduction to Infrared and Raman Spectroscopy,* 3rd ed., Academic Press, Boston, 1990.
2. Conley, R.T., *Infrared Spectroscopy,* Allyn and Bacon, Boston, 1972.

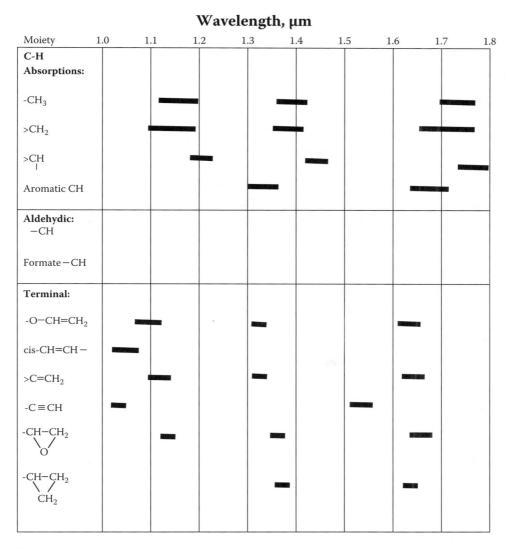

Figure 2.12

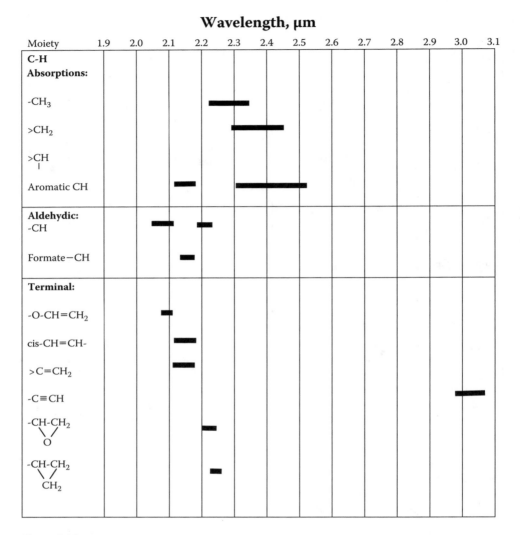

Figure 2.13

Wavelength, µm

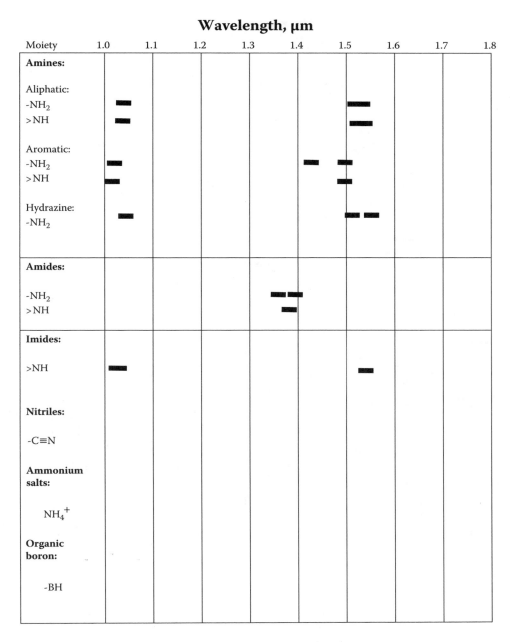

Figure 2.14

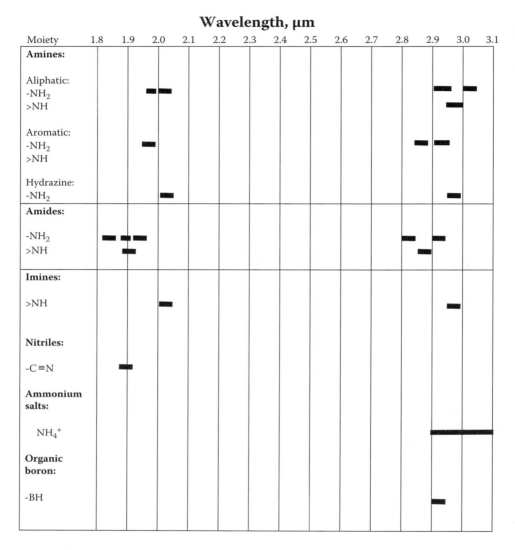

Figure 2.15

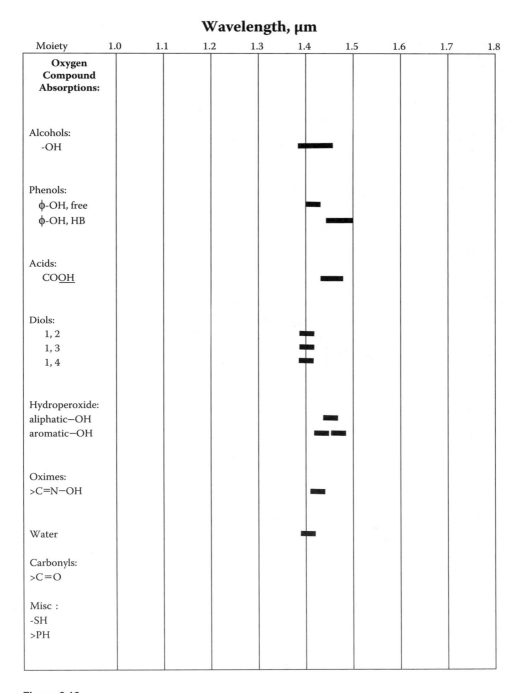

Figure 2.16

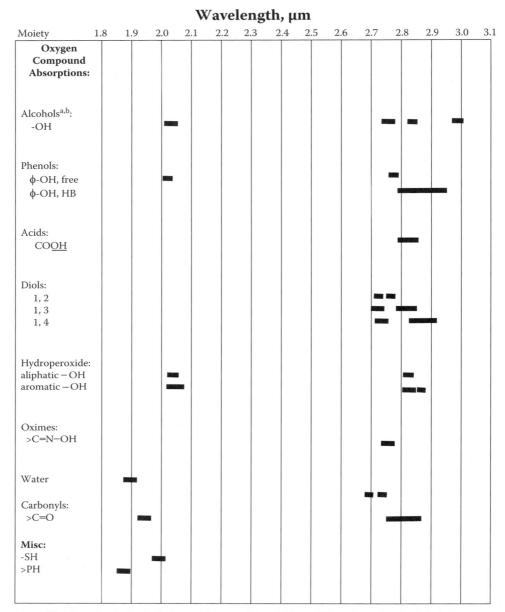

Notes: [a] The dimer bands for alcohols occur between 2.8 and 2.9 µm; [b] The polymer bands of alcohols occur between 2.96 and 3.05 µm; HB designates the presence of hydrogen bonding.

Figure 2.17

INORGANIC GROUP ABSORPTIONS

The following chart provides the infrared absorbance that can be observed from inorganic functional moieties. These have been compiled from a study of the IR absorption of a number of inorganic species.[1] It should be understood that the physical state of the sample plays a role in the intensity and position of these bands. These variables include crystal structure, crystallite size, water of hydration, etc. This chart must therefore be regarded as an approximate guide.

REFERENCE

1. Miller, F.A. and Wilkins, C.H., Infrared spectra and characteristic frequencies of inorganic ions, *Anal. Chem.*, 24, 1253–94, 1952.

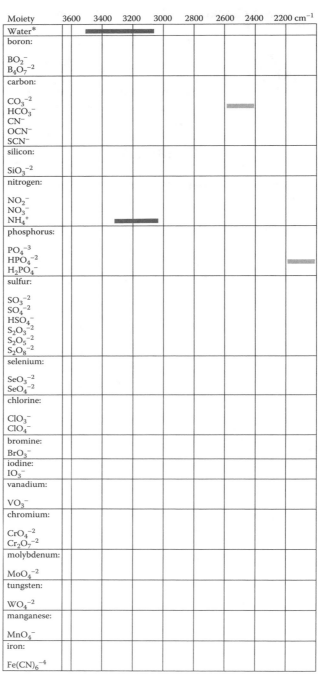

Moiety	3600	3400	3200	3000	2800	2600	2400	2200 cm^{-1}
Water*		■■■■■■■■■ (strong)						
boron:								
BO_2^-								
$B_4O_7^{-2}$								
carbon:								
CO_3^{-2}								
HCO_3^-						■■■ (medium)		
CN^-								
OCN^-								
SCN^-								
silicon:								
SiO_3^{-2}								
nitrogen:								
NO_2^-								
NO_3^-								
NH_4^+			■■■■■ (strong)					
phosphorus:								
PO_4^{-3}								
HPO_4^{-2}								
$H_2PO_4^-$								■■ (weak)
sulfur:								
SO_3^{-2}								
SO_4^{-2}								
HSO_4^-								
$S_2O_3^{-2}$								
$S_2O_5^{-2}$								
$S_2O_8^{-2}$								
selenium:								
SeO_3^{-2}								
SeO_4^{-2}								
chlorine:								
ClO_3^-								
ClO_4^-								
bromine:								
BrO_3^-								
iodine:								
IO_3^-								
vanadium:								
VO_3^-								
chromium:								
CrO_4^{-2}								
$Cr_2O_7^{-2}$								
molybdenum:								
MoO_4^{-2}								
tungsten:								
WO_4^{-2}								
manganese:								
MnO_4^-								
iron:								
$Fe(CN)_6^{-4}$								

* This water is water of crystallization

Figure 2.18 Key for level of infrared absorbance:
Strong: ■■■■■■
Medium: ■■■■■■
Weak: ■■■■■■

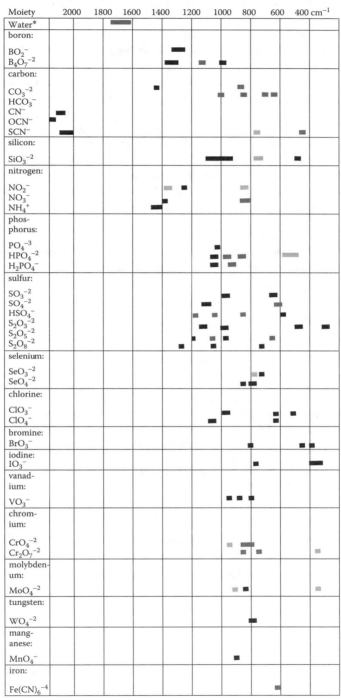

Figure 2.19 Key for level of infrared absorbance:
Strong:
Medium:
Weak:

USEFUL SOLVENTS FOR INFRARED SPECTROPHOTOMETRY

The following tables provide the infrared absorption spectra of several useful solvents, along with solvent design properties.[1–10] In most cases, two spectra are provided for each solvent. The first in each set was measured using a double-beam spectrophotometer using a neat sample against an air reference. These spectra are presented in both wavenumber (cm^{-1}) and micrometer (μm) scales. The spectra were recorded under high-concentration conditions (in terms of path length and attenuation) to emphasize the characteristics of each solvent. Thus, these spectra are not meant to be "textbook" examples of infrared spectra. The second spectrum in each set was measured with a Fourier transform instrument. The physical properties listed are those needed most often in designing spectrophotometric experiments.[1–10] The refractive indices are values measured with the sodium-d line. Solvation properties include the solubility parameter, δ, hydrogen bond index, λ, and the solvatochromic parameters α, β, and π^*. Please note that the Chemical Abstract Service registry numbers are also provided for each solvent, to allow the reader to easily obtain further information using computerized database services. Note that the heat of vaporization is presented in the commonly used cal/g unit. To convert to the appropriate SI unit (J/g), multiply by 4.184.

We realize that a number of the solvents listed here are not permitted in some academic laboratories. Information on these solvents is presented for users in laboratories equipped to deal with the hazards associated with them.

REFERENCES

1. Lewis, R.J., *Hawley's Condensed Chemical Dictionary,* 14th ed., John Wiley and Sons, New York, 2002.
2. Dreisbach, R.R., *Physical Properties of Chemical Compounds*, in Advances in Chemistry Series, No. 22, American Chemical Society, Washington, DC, 1959.
3. Jamieson, D.T., Irving, J.B., and Tudhope, J.S., *Liquid Thermal Conductivity: A Data Survey to 1973,* Her Majesty's Stationery Office, Edinburgh, 1975.
4. Lewis, R.J. and Sax, N.I., *Sax's Dangerous Properties of Industrial Materials,* 9th ed., Thompson Publishing, Washington, DC, 1995.
5. Sedivec, V. and Flek, J., *Handbook of Analysis of Organic Solvents,* John Wiley and Sons (Halsted Press), New York, 1976.
6. Epstein, W.W. and Sweat, F.W., Dimethyl sulfoxide oxidations, *Chem. Rev.* 67(3): 247–260, 1967.
7. Lide, D.R., Ed., *CRC Handbook for Chemistry and Physics,* 83rd ed., CRC Press, Boca Raton, Fl, 2003.
8. Bruno, T.J. and Svoronos, P.D.N., *CRC Handbook of Basic Tables for Chemical Analysis,* 2nd ed., CRC Press, Boca Raton, FL, 2003.
9. NIST, NIST Chemistry Web Book, NIST Standard Reference Database, No. 69, March 2003. http://webbook.nist.gov/chemistry.
10. Marcus, Y., The properties of organic liquids that are relevant to their use as solvating solvents, *Chem. Soc. Rev.,* 22(6): 409–416, 1993.

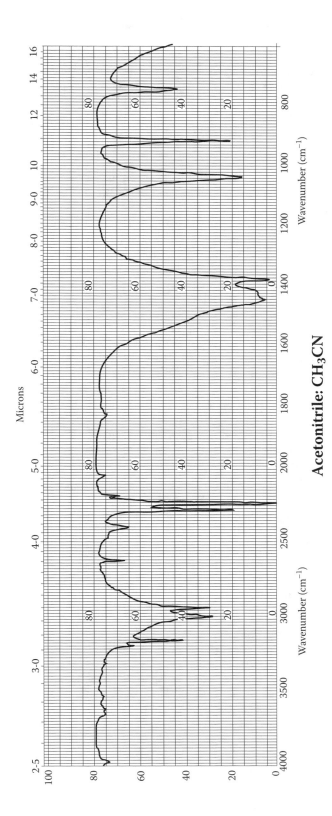

Acetonitrile: CH₃CN

Figure 2.20

ACETONITRILE, CH₃CN

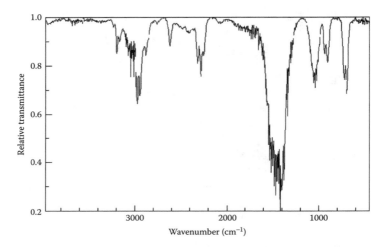

Figure 2.21

Physical Properties

Relative molecular mass	41.05
Melting point	−45.7°C
Normal boiling point	81.6°C
Refractive index (20°C)	1.34423
Density (20°C)	0.7857 g/mL
Viscosity (25°C)	0.345 mPa · s
Surface tension (20°C)	29.30 mN/m
Heat of vaporization (at boiling point)	29.75 kJ/mol
Thermal conductivity (20°C)	0.1762 W/(m · K)
Dielectric constant (20°C)	38.8
Relative vapor density (air = 1)	1.41
Vapor pressure (20°C)	0.0097 MPa
Solubility in water[a]	∞
Flash point (OC)	6°C
Autoignition temperature	509°C
Explosive limits in air	4.4–16%, vol/vol
CAS registry number	75-05-8
Exposure limits	40 ppm, 8-hr TWA
Solubility parameter, δ	11.9
Solvatochromic α	0.19
Solvatochromic β	0.4
Solvatochromic π*	0.75

[a] Forms azeotrope with water (at 16% mass/mass) that boils at 76°C.

Notes: Highly polar solvent; sweet, ethereal odor; soluble in water; flammable, burns with a luminous flame; highly toxic by ingestion, inhalation, and skin absorption; miscible with water, methanol, methyl acetate, ethyl acetate, acetone, ethers, acetamide solutions, chloroform, carbon tetrachloride, ethylene chloride, and many unsaturated hydrocarbons; immiscible with many saturated hydrocarbons (petroleum fractions); dissolves some inorganic salts such as silver nitrate, lithium nitrate, magnesium bromide; incompatible with strong oxidants; hydrolyzes in the presence of aqueous bases and strong aqueous acids.

Synonyms: methyl cyanide, acetic acid nitrile, cyanomethane, ethylnitrile.

BENZENE C_6H_6

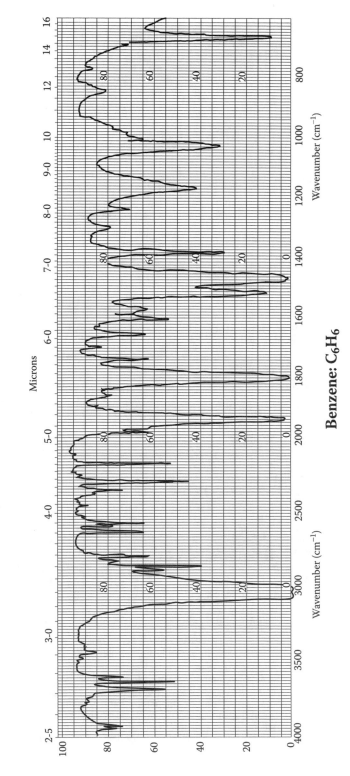

Benzene: C_6H_6

Figure 2.22

BENZENE, C_6H_6

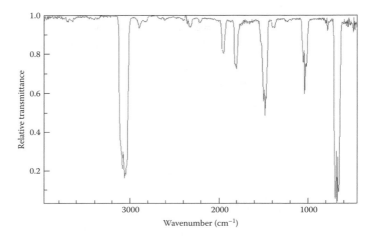

Figure 2.23

Physical Properties

Relative molecular mass	78.11
Melting point	5.5°C
Normal boiling point	80.1°C
Refractive index	
(20°C)	1.50110
(25°C)	1.4979
Density	
(20°C)	0.8790 g/mL
(25°C)	0.8737 g/mL
Viscosity (25°C)	0.654 mPa·s
Surface tension (20°C)	28.87 mN/m
Heat of vaporization (at boiling point)	30.72 kJ/mol
Thermal conductivity (25°C)	0.1424 W/(m·K)
Dielectric constant (20°C)	2.284
Relative vapor density (air = 1)	2.77
Vapor pressure (25°C)	0.0097 MPa
Solubility in water[a]	0.07%, mass/mass
Flash point (OC)	−11°C
Autoignition temperature	562°C
Explosive limits in air	1.4–8.0%, vol/vol
CAS registry number	71-43-2
Exposure limits	10 ppm, 8-h TWA
Solubility parameter, δ	9.2
Hydrogen bond index, λ	2.2
Solvatochromic α	0.00
Solvatochromic β	0.10
Solvatochromic π^*	0.59

[a] Forms azeotrope with ethanol (approximately 65°C).

Notes: Nonpolar, aromatic solvent; sweet odor; very flammable and toxic; confirmed human carcinogen; soluble in alcohols, hydrocarbons (aliphatic and aromatic), ether, chloroform, carbon tetrachloride, carbon disulfide, slightly soluble in water. Incompatible with some strong acids and oxidants, chlorine trifluoride/zinc (in the presence of steam); dimerizes at high temperature to form biphenyl. **Confirmed human carcinogen**.

Synonyms: cyclohexatriene, benzin, benzol, phenylhydride. These are the most common, although there are many other synonyms.

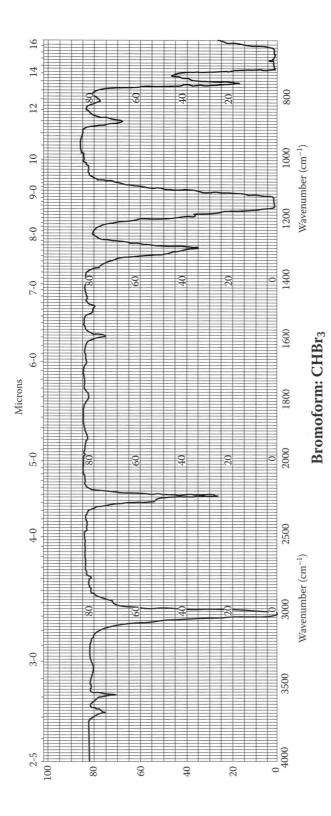

Bromoform: CHBr₃

Figure 2.24

BROMOFORM, CHBr$_3$

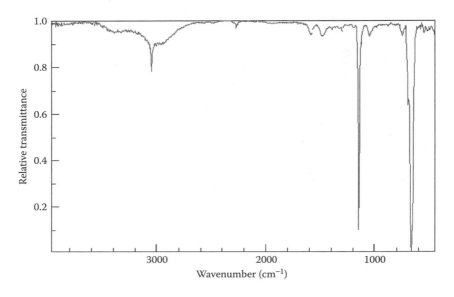

Figure 2.25

Physical Properties

Relative molecular mass	252.75
Melting point	5.7°C
Normal boiling point	149.5°C
Refractive index (20°C)	1.6005
Density (20°C)	2.8899 g/mL
Viscosity (25°C)	1.89 mPa·s
Surface tension (20°C)	41.53 mN/m
Heat of vaporization (at boiling point)	39.66 kJ/mol
Thermal conductivity (20°C)	0.0961 W/(m·K)
Dielectric constant (20°C)	4.39
Relative vapor density (air = 1)	2.77
Vapor pressure (25°C)	0.0008 MPa
Solubility in water	Slightly
Flash point (OC)	Nonflammable
Autoignition temperature	Not determined
Explosive limits in air	Nonflammable
CAS registry number	75-25-2
Exposure limits	0.5 ppm (skin)
Solvatochromic α	0.05
Solvatochromic β	0.05
Solvatochromic π^*	0.62

Notes: Moderately polar, weakly hydrogen-bonding solvent, dense liquid; gradually decomposes to acquire a yellow color, with air and/or light accelerating this decomposition; nonflammable; commercial product is often stabilized by the addition of 3–4% (mass/mass) alcohols; highly toxic by ingestion, inhalation, and skin absorption; soluble in alcohols, organohalogen compounds, hydrocarbons, benzene, and many oils. Incompatible with many alkali and alkaline earth metals.

Synonyms: tribromomethane.

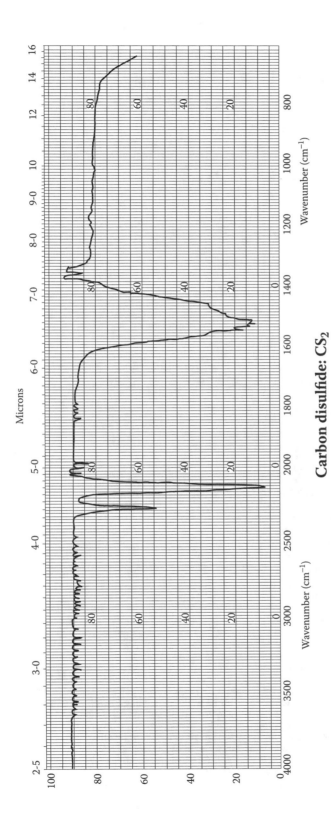

Carbon disulfide: CS$_2$

Figure 2.26

CARBON DISULFIDE, CS₂

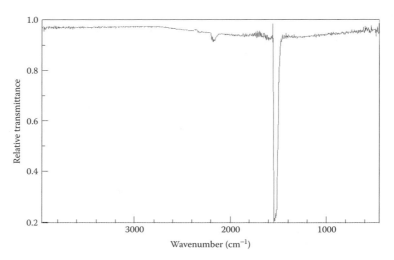

Figure 2.27

Physical Properties

Relative molecular mass	76.14
Melting point	−111°C
Normal boiling point	46.3°C
Refractive index	
(20°C)	1.6280
(25°C)	1.6232
Density	
(20°C)	1.2631 g/mL
(25°C)	1.2556 g/mL
Viscosity (20°C)	0.363 mPa·s
Surface tension (20°C)	32.25 mN/m
Heat of vaporization (at boiling point)	26.74 kJ/mol
Dielectric constant (20°C)	2.641
Relative vapor density (air = 1)	2.64
Vapor pressure (25°C)	0.0448 MPa
Solubility in water (20°C)	0.29%, mass/mass
Flash point (OC)	−30°C
Autoignition temperature	100°C
Explosive limits in air	1.0–50%, vol/vol
CAS registry number	75-15-0
Exposure limits	20 ppm, 8-hr TWA
Solvatochromic α	0.00
Solvatochromic β	0.07
Solvatochromic π*	0.61

Notes: Moderately polar solvent, soluble in alcohols, benzene, ethers, and chloroform; slightly soluble in water; very flammable and mobile; can be ignited by friction or contact with hot surfaces such as steam pipes; burns with a blue flame to produce carbon dioxide and sulfur dioxide; toxic by inhalation, ingestion, and skin absorption; strong disagreeable odor when impure; incompatible with aluminum (powder), azides, chlorine, chlorine monoxide, ethylene diamine, ethyleneamine, fluorine, nitrogen oxides, potassium, and zinc and other oxidants; soluble in methanol, ethanol, ethers, benzene, chloroform, carbon tetrachloride, and many oils; can be stored in metal, glass porcelain, and Teflon containers.

Synonyms: carbon bisulfide, dithiocarbon anhydride.

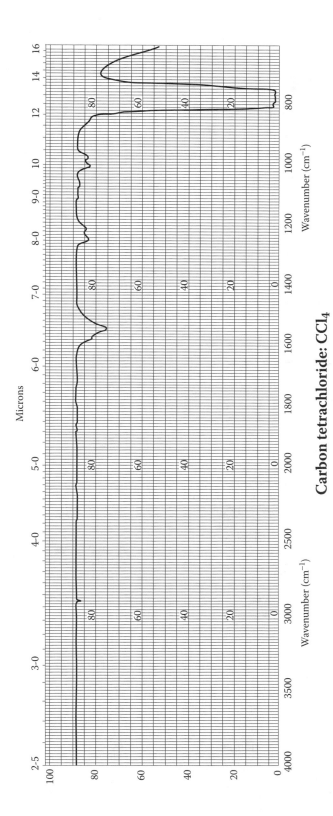

Carbon tetrachloride: CCl$_4$

Figure 2.28

CARBON TETRACHLORIDE, CCl$_4$

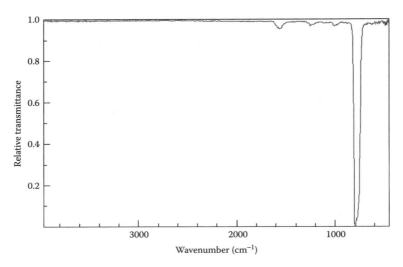

Figure 2.29

Physical Properties

Relative molecular mass	153.82
Melting point	−22.85°C
Normal boiling point	76.65°C
Refractive index	
(20°C)	1.4607
(25°C)	1.4570
Density:	
(20°C)	1.5940 g/mL
(25°C)	1.5843 g/mL
Viscosity (20°C)	0.969 mPa · s
Surface tension (20°C)	26.75 mN/m
Heat of vaporization (at boiling point)	29.82 kJ/mol
Thermal conductivity (20°C)	0.1070 W/(m · K)
Dielectric constant (20°C)	2.238
Relative vapor density (air = 1)	5.32
Vapor pressure (25°C)	0.0122 MPa
Solubility in water (20°C)	0.08, w/w
Flash point (OC)	Noncombustible
Autoignition temperature	Noncombustible
Explosive limits in air	Nonexplosive
CAS registry number	56-23-5
Exposure limits	5 ppm (skin)
Solubility parameter, δ	8.6
Hydrogen bond index, λ	2.2
Solvatochromic α	0.00
Solvatochromic β	0.10
Solvatochromic π*	0.28

Notes: Nonpolar solvent; soluble in alcohols, ethers, chloroform and other halocarbons, benzene, and most fixed and volatile oils; insoluble in water; nonflammable; extremely toxic by inhalation, ingestion, or skin absorption; carcinogenic; incompatible with allyl alcohol, silanes, triethyldialuminum, many metals (e.g., sodium).

Synonyms: tetrachloromethane, perchloromethane, methane tetrachloride, Halon-104.

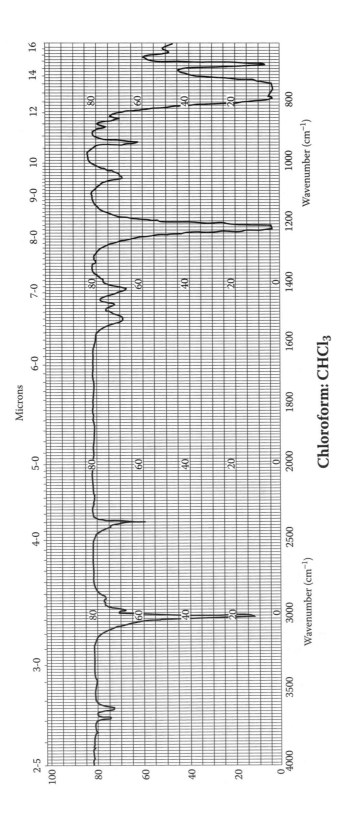

Chloroform: CHCl₃

Figure 2.30

CHLOROFORM, CHCl$_3$

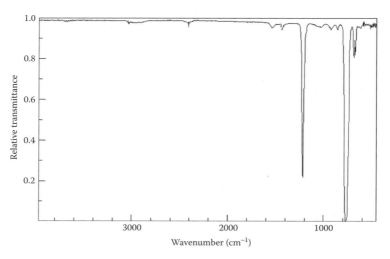

Figure 2.31

Physical Properties

Relative molecular mass	119.38
Melting point	−63.2°C
Normal boiling point	61.2°C
Refractive index	
(20°C)	1.4458
(25°C)	1.4422
Density	
(20°C)	1.4892 g/mL
(25°C)	1.4798 g/mL
Viscosity (20°C)	0.566 mPa · s
Surface tension	27.2 mN/m
Heat of vaporization (at boiling point)	29.24 kJ/mol
Thermal conductivity (20°C)	0.1164 W/(m · K)
Dielectric constant (20°C)	4.806
Relative vapor density (air = 1)	4.13
Vapor pressure (25°C)	0.0263 MPa
Solubility in water	0.815%, w/w
Flash point (OC)	Noncombustible[a]
Autoignition temperature	Noncombustible[a]
Explosive limits in air	Nonexplosive
CAS registry number	67-66-3
Exposure limits	10 ppm, 8-hr TWA
Solubility parameter, δ	9.3
Hydrogen bond index, λ	2.2
Solvatochromic α	0.20
Solvatochromic β	0.10
Solvatochromic π*	0.58

[a] Although chloroform is nonflammable, it will burn upon prolonged exposure to flame or high temperature.

Notes: Polar solvent; soluble in alcohols, ether, benzene, and most oils; usually stabilized with methanol to prevent phosgene formation; flammable and highly toxic by inhalation, ingestion, or skin absorption; narcotic; suspected to be carcinogenic; incompatible with caustics, active metals, aluminum powder, potassium, sodium, magnesium.

Synonyms: trichloromethane, methane trichloride.

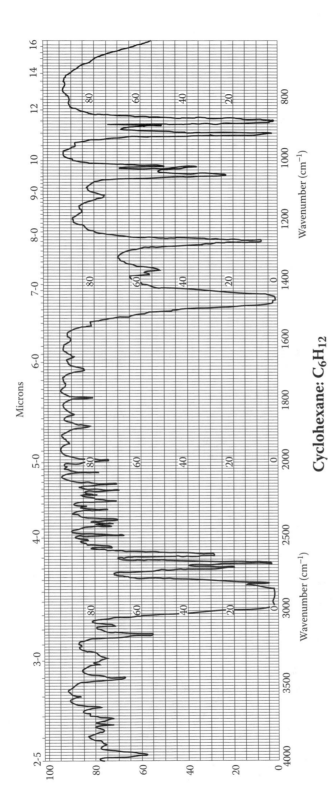

Cyclohexane: C$_6$H$_{12}$

Figure 2.32

CYCLOHEXANE, C_6H_{12}

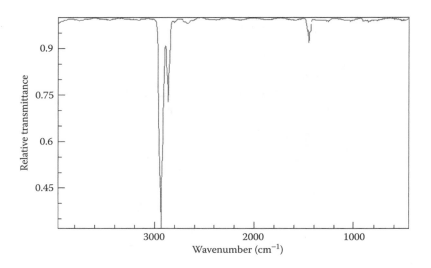

Figure 2.33

Physical Properties

Relative molecular mass	84.16
Melting point	6.3°C
Normal boiling point	80.7°C
Refractive index	
(20°C)	1.4263
(25°C)	1.4235
Density	
(20°C)	0.7786 g/mL
(25°C)	0.7739 g/mL
Viscosity (20°C)	1.06 mPa·s
Surface tension (20°C)	24.99 mN/m
Heat of vaporization (at boiling point)	29.97 kJ/mol
Thermal conductivity (20°C)	0.122 W/(m·K)
Dielectric constant (20°C)	2.023
Relative vapor density (air = 1)	2.90
Vapor pressure (25°C)	0.0111 MPa
Solubility in water (20°C)	<0.01%, mass/mass
Flash point (OC)	−17°C
Autoignition temperature	245°C
Explosive limits in air	1.31–8.35%, vol/vol
CAS registry number	110-82-7
Exposure limits	330 ppm, 8-hr TWA
Solvatochromic α	0.00
Solvatochromic β	0.00
Solvatochromic π*	0.00

Notes: Nonpolar hydrocarbon solvent; mild, gasoline-like odor; soluble in hydrocarbons, alcohols, organic halides, acetone, benzene; flammable; moderately toxic by inhalation, ingestion, or skin absorption; may be narcotic at high concentrations; reacts with oxygen (air) at elevated temperatures; decomposes upon heating; incompatible with strong oxidants.

Synonyms: benzene hexahydride, hexamethylene, hexanaphthene, hexahydrobenzene.

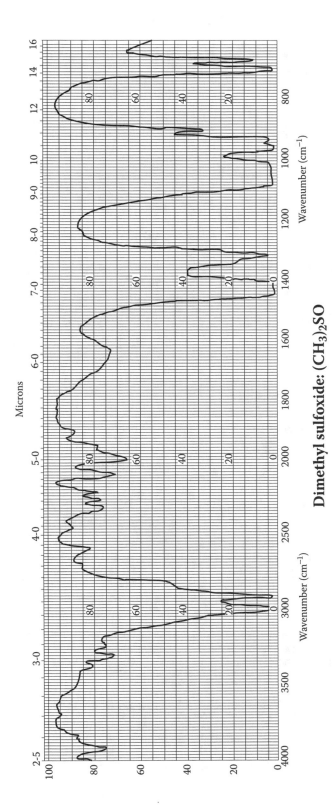

Dimethyl sulfoxide: (CH$_3$)$_2$SO

Figure 2.34

DIMETHYL SULFOXIDE, $(CH_3)_2SO$

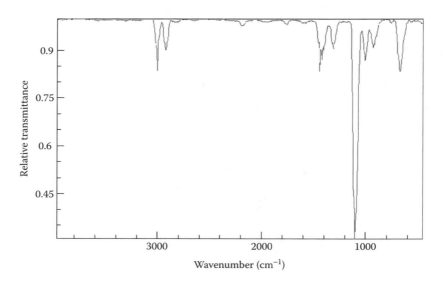

Figure 2.35

Physical Properties

Relative molecular mass	78.13
Melting point	18.5°C
Normal boiling point	189°C
Refractive index (20°C)	1.4770
Density (20°C)	1.1014 g/mL
Viscosity (25°C)	1.98 mPa·s
Surface tension	43.5 mN/m
Relative vapor density (air = 1)	2.7
Vapor pressure	5.3×10^{-5} MPa
Solubility in water	∞
Flash point (OC)	95°C
Autoignition temperature	215°C
Explosive limits in air	26.0–28.5%, vol/vol
CAS registry number	67-68-5
Exposure limits	None established
Solubility parameter, δ	13.0
Hydrogen bond index, λ	5.0
Solvatochromic α	0.00
Solvatochromic β	0.76
Solvatochromic π^*	1.00

Notes: Colorless, odorless (when pure), hygroscopic liquid; powerful aprotic solvent; dissolves many inorganic salts, soluble in water; combustible; readily penetrates the skin; incompatible with strong oxidizers and many halogenated compounds (e.g., alkyl halides, aryl halides), oxygen, peroxides, diborane, perchlorates.

Synonyms: DMSO, methyl sulfoxide, sulfinylbismethane.

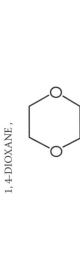

1, 4-DIOXANE,

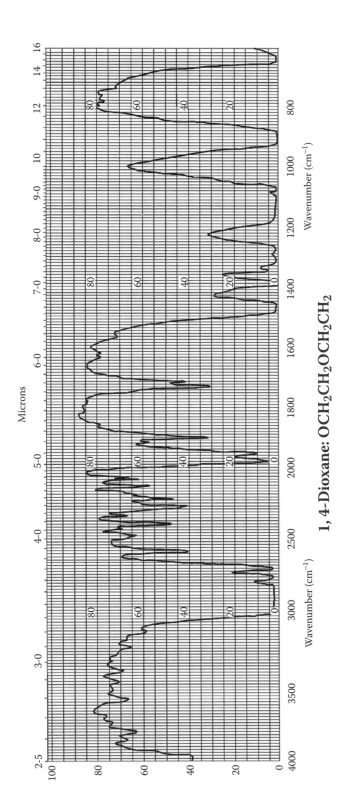

1, 4-Dioxane: OCH₂CH₂OCH₂CH₂

Figure 2.36

1,4-DIOXANE, OCH$_2$CH$_2$OCH$_2$CH$_2$

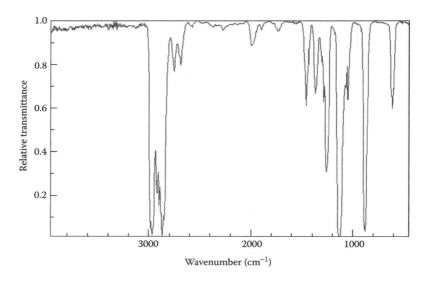

Figure 2.37

Physical Properties

Relative molecular mass	88.11
Melting point	11°C
Boiling point	101.3°C
Refractive index	
(20°C)	1.4221
(25°C)	1.4195
Density	
(20°C)	1.0338 g/mL
(25°C)	1.0282 g/mL
Viscosity (20°C)	1.37 mPa·s
Surface tension (20°C)	33.74 mN/m
Heat of vaporization (at boiling point)	34.16 kJ/mol
Dielectric constant (20°C)	2.209
Relative vapor density (air = 1)	3.03
Vapor pressure (25°C)	0.0053 MPa
Solubility in water	∞
Flash point (OC)	12°C
Autoignition temperature	180°C
Explosive limits in air	1.97–22.2%, vol/vol
CAS registry number	123-91-1
Exposure limits	100 ppm (skin)
Solubility parameter, δ	9.9
Hydrogen bond index, λ	5.7
Solvatochromic α	0.00
Solvatochromic β	0.37
Solvatochromic π*	0.55

Notes: Moderately polar solvent; soluble in water and most organic solvents; flammable; highly toxic by ingestion and inhalation; absorbed through the skin; may cause central nervous system depression and necrosis of the liver and kidneys; incompatible with strong oxidizers.

Synonyms: diethylene ether, 1,4-diethylene dioxide, diethylene dioxide, dioxyethylene ether.

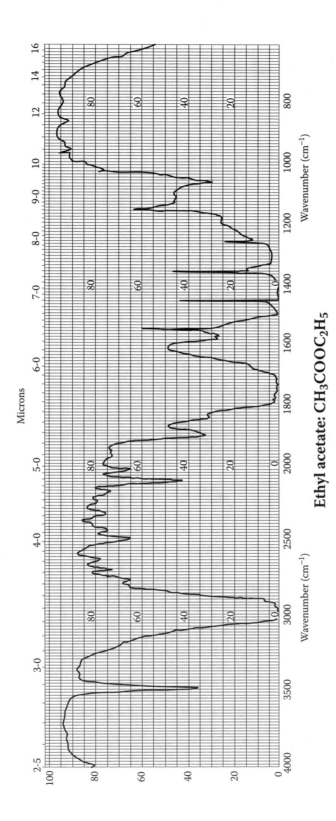

Ethyl acetate: CH₃COOC₂H₅

Figure 2.38

ETHYL ACETATE, $CH_3COOC_2H_5$

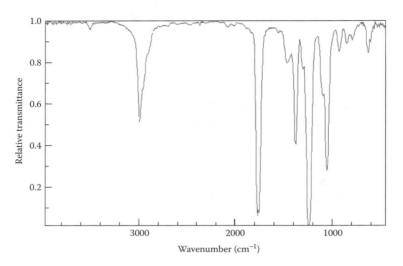

Figure 2.39

Physical Properties

Relative molecular mass	88.11
Melting point	−83.58°C
Boiling point	77.06°C
Refractive index	
(20°C)	1.3723
(25°C)	1.3698
Density	
(20°C)	0.9006 g/mL
(25°C)	0.8946 g/mL
Viscosity (20°C)	0.452 mPa·s
Surface tension (20°C)	23.95 mN/m
Heat of vaporization (at boiling point)	31.94 kJ/mol
Thermal conductivity (20°C)	0.122 W/(m·K)
Dielectric constant (25°C)	6.02
Relative vapor density (air = 1)	3.04
Vapor pressure (20°C)	0.0097 MPa
Solubility in water (20°C)[a]	3.3%, mass/mass
Flash point (OC)	−1°C
Autoignition temperature	486°C
Explosive limits in air	2.18–11.5%, vol/vol
CAS registry number	141-78-6
Exposure limits	440 ppm, 8-hr TWA
Solubility parameter, δ	9.1
Hydrogen bond index, λ	5.2
Solvatochromic α	0.00
Solvatochromic β	0.43
Solvatochromic π^*	0.55

[a] Forms an azeotrope with water at 6.1%, mass/mass, that boils at 70.4°C.

Notes: Polar solvent; insoluble in water, soluble in alcohols, organic halides, ether, and many oils; flammable; moderately toxic by inhalation and skin absorption; incompatible with strong oxidizers, nitrates, strong alkalis, strong acids.

Synonyms: acedin, acetic ether, acetic ester, vinegar naphtha, acetic acid ethyl ester.

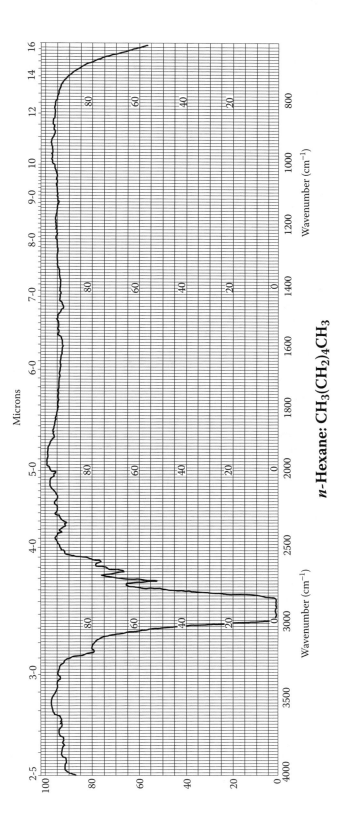

n-Hexane: CH₃(CH₂)₄CH₃

Figure 2.40

n-HEXANE, CH$_3$(CH$_2$)$_4$CH$_3$

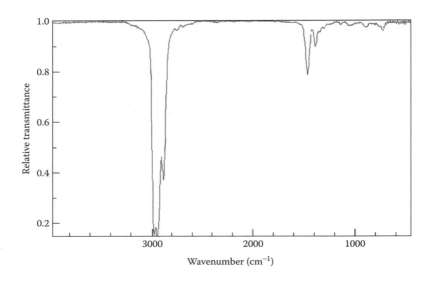

Figure 2.41

Physical Properties

Relative molecular mass	86.18
Melting point	−95°C
Normal boiling point	68.742°C
Refractive index	
(20°C)	1.37486
(25°C)	1.3723
Density	
(20°C)	0.6594 g/mL
(25°C)	0.6548 g/mL
Viscosity (20°C)	0.31 mPa·s
Surface tension (20°C)	18.42 mN/m
Heat of vaporization (at boiling point)	28.85 kJ/mol
Thermal conductivity (20°C)	0.1217 W/(m·K)
Dielectric constant (20°C)	1.890
Relative vapor density (air = 1)	2.97
Vapor pressure (25°C)	0.0222 MPa
Solubility in water (20°C)	0.011%, mass/mass
Flash point (OC)	−26°C
Autoignition temperature	247°C
Explosive limits in air	1.25–6.90%, vol/vol
CAS registry number	110-54-3
Exposure limits	500 ppm, 8-hr TWA
Solubility parameter, δ	9.3
Hydrogen bond index, λ	2.2
Solvatochromic α	0.00
Solvatochromic β	0.00
Solvatochromic π*	0.08

Notes: Nonpolar solvent; soluble in alcohols, hydrocarbons, organic halides, acetone, and ethers; insoluble in water; flammable; moderately toxic by inhalation and ingestion; incompatible with strong oxidizers.

Synonyms: hexane, hexyl hydride.

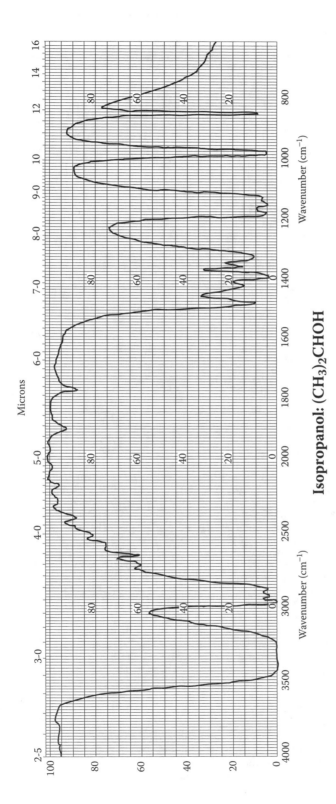

Isopropanol: (CH₃)₂CHOH

Figure 2.42

ISOPROPANOL, (CH₃)₂CHOH

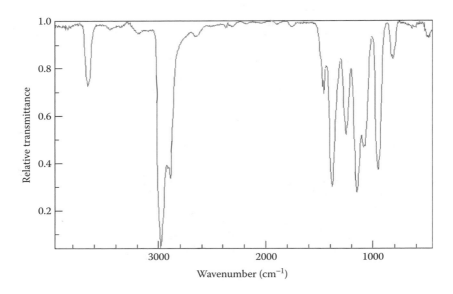

Figure 2.43

Physical Properties

Relative molecular mass	60.10
Melting point	89.8°C
Boiling point	82.4°C
Refractive index	
(20°C)	1.3771
(25°C)	1.3750
Density	
(20°C)	0.7864 g/mL
(25°C)	0.7812 g/mL
Viscosity (20°C)	2.43 mPa·s
Surface tension (20°C)	21.99 mN/m
Heat of vaporization (at boiling point)	39.85 kJ/mol
Dielectric constant (25°C)	18.3
Relative vapor density (air = 1)	2.07
Vapor pressure	0.0044 MPa
Solubility in water (20°C)	∞
Flash point (OC)	16°C
Autoignition temperature	456°C
Explosive limits in air	2.02–11.8%, vol/vol
CAS registry number	67-63-0
Exposure limits	400 ppm (skin)
Solubility parameter, δ	11.5
Hydrogen bond index, λ	8.9
Solvatochromic α	0.76
Solvatochromic β	0.84
Solvatochromic π^*	0.48

Notes: Polar solvent; soluble in water, alcohols, ethers, and many hydrocarbons and oils; flammable and moderately toxic by ingestion, inhalation, and skin absorption; incompatible with strong oxidizers.

Synonyms: dimethyl carbinol, sec-propyl alcohol, 2-propanol, isopropyl alcohol.

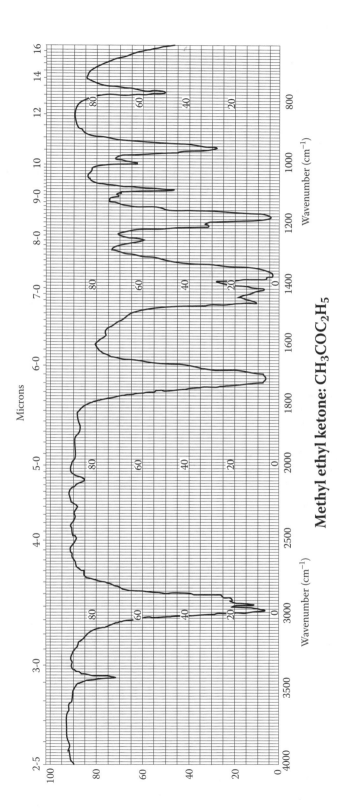

Methyl ethyl ketone: CH₃COC₂H₅

Figure 2.44

METHYL ETHYL KETONE, CH₃COC₂H₅

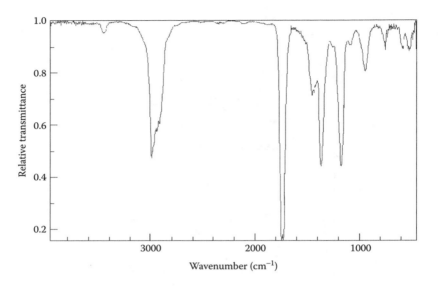

Figure 2.45

Physical Properties

Relative molecular mass	72.11
Melting point	−86.4°C
Boiling point	79.6°C
Refractive index	
(20°C)	1.379
(25°C)	1.3761
Density	
(20°C)	0.8054 g/mL
(25°C)	0.8002 g/mL
Viscosity (20°C)	0.448 mPa·s
Surface tension (20°C)	24.50 mN/m
Heat of vaporization (at boiling point)	31.3 kJ/mol
Thermal conductivity (20°C)	0.1465 W/(m·K)
Dielectric constant (20°C)	18.5
Relative vapor density (air = 1)	2.41
Vapor pressure (25°C)	0.0129 MPa
Solubility in water (20°C)	27.33%, mass/mass
Flash point (OC)	2°C
Autoignition temperature	516°C
Explosive limits in air	1.81–11.5%, vol/vol
CAS registry number	78-93-3
Exposure limits	200 ppm, 8-hr TWA
Solubility parameter, δ	9.3
Hydrogen bond index, λ	5.0
Solvatochromic α	0.6
Solvatochromic β	0.48
Solvatochromic π^*	0.67

Notes: Polar solvent; soluble in water, ketones, organic halides, alcohols, ether, and many oils; highly flammable; narcotic by inhalation; incompatible with strong oxidizers, nitrates, nitric acid, reducing agents.

Synonyms: ethyl methyl ketone, 2-butanone, methyl acetone, MEK.

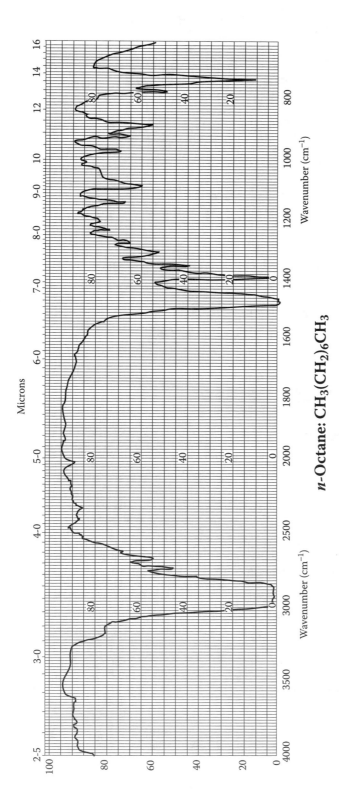

n-Octane: CH₃(CH₂)₆CH₃

Figure 2.46

n-OCTANE, CH₃(CH₂)₆CH₃

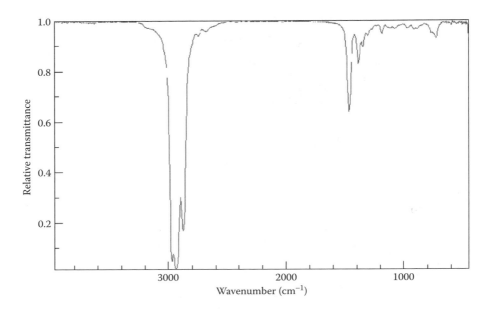

Figure 2.47

Physical Properties

Relative molecular mass	114.23
Melting point	−56.7°C
Boiling point	125.6°C
Refractive index	
(20°C)	1.39745
(25°C)	1.3951
Density	
(20°C)	0.7025 g/mL
(25°C)	0.6985 g/mL
Viscosity (20°C)	0.539 mPa · s
Surface tension (20°C)	21.75 mN/m
Heat of vaporization (at boiling point)	34.41 kJ/mol
Dielectric constant (20°C)	1.948
Relative vapor density (air = 1)	3.86
Vapor pressure (25°C)	0.0023 MPa
Solubility in water (20°C)	≈0.002%, mass/mass
Flash point (CC)	13°C
Autoignition temperature	232°C
Explosive limits in air	0.84–3.2%, vol/vol
CAS registry number	111-65-9
Exposure limits	550 ppm, 8-hr TWA
Hydrogen bond index, λ	2.2
Solvatochromic α	0.00
Solvatochromic β	0.00
Solvatochromic π*	0.01

Notes: Nonpolar solvent; soluble in alcohol, acetone, and hydrocarbons; insoluble in water; flammable; incompatible with strong oxidizers.

Synonyms: octane.

TETRAHYDROFURAN

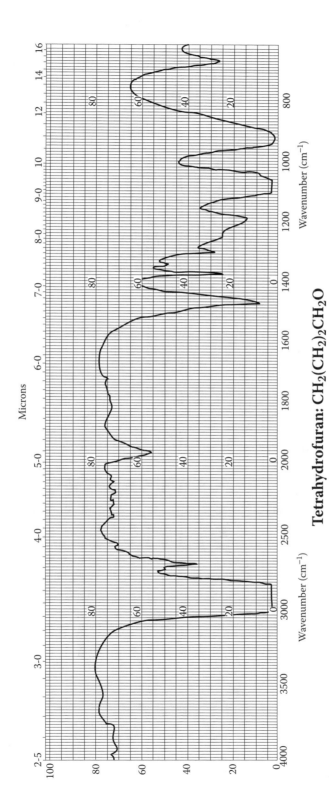

Tetrahydrofuran: CH$_2$(CH$_2$)$_2$CH$_2$O

Figure 2.48

TETRAHYDROFURAN, $CH_2(CH_2)_2CH_2O$

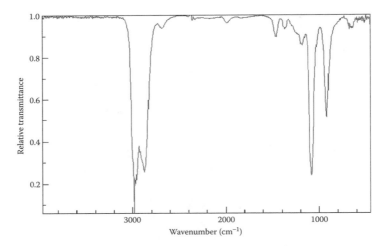

Figure 2.49

Physical Properties

Relative molecular mass	72.108
Melting point	−65°C
Normal boiling point	66°C
Refractive index	
(20°C)	1.4070
(25°C)	1.4040
Density	
(20°C)	0.8880 g/mL
(25°C)	0.8818 g/mL
Viscosity (20°C)	0.55 mPa·s
Surface tension (20°C)	26.4 mN/m
Heat of vaporization (at boiling point)	29.81 kJ/mol
Dielectric constant (20°C)	7.54
Relative vapor density (air = 1)	2.5
Vapor pressure (20°C)	0.0191 MPa
Solubility in water (20°C)[a]	∞
Flash point (CC)	−17°C
Autoignition temperature	260°C
Explosive limits in air	1.8–11.8%, vol/vol
CAS registry number	109-99-9
Exposure limits	200 ppm, 8-hr TWA
Solubility parameter, δ	9.1
Hydrogen bond index, λ	5.3
Solvatochromic α	0.00
Solvatochromic β	0.55
Solvatochromic π*	0.58

[a] pH of aqueous solution = 7.

Notes: Moderately polar solvent, ethereal odor; soluble in water and most organic solvents; flammable; moderately toxic; incompatible with strong oxidizers; can form potentially explosive peroxides upon long standing in air; commercially, it is often stabilized against peroxidation with 0.5–1.0% (mass/mass) p-cresol, 0.05 −1.0% (mass/mass) hydroquinone, or 0.01% (mass/mass) 4,4′-thiobis(6-tert-butyl-m-cresol); can polymerize in the presence of cationic initiators such as Lewis acids or strong proton acids.

Synonyms: THF, tetramethylene oxide, diethylene oxide, 1,4-epoxybutane oxolane, oxacyclopentane.

TOLUENE CH₃C₆H₅

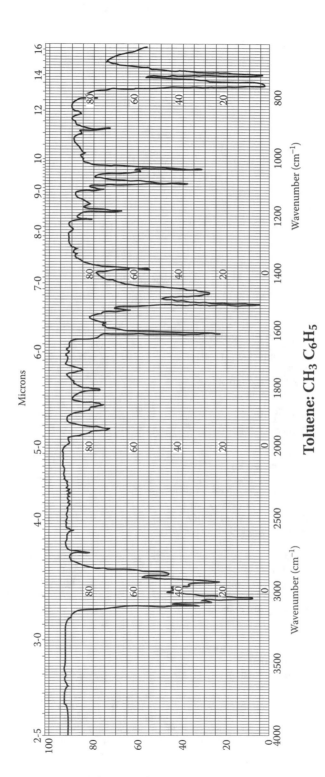

Toluene: CH₃ C₆H₅

Figure 2.50

TOLUENE, $CH_3C_6H_5$

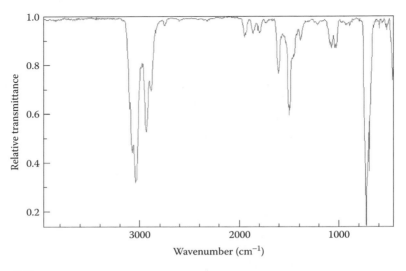

Figure 2.51

Physical Properties

Relative molecular mass	92.14
Melting point	−94.5°C
Normal boiling point	110.7°C
Refractive index	
(20°C)	1.497
(25°C)	1.4941
Density	
(20°C)	0.8669 g/mL
(25°C)	0.8623 g/mL
Viscosity (20°C)	0.587 mPa·s
Surface tension (20°C)	28.52 mN/m
Heat of vaporization (at boiling point)	33.18 kJ/mol
Thermal conductivity (20°C)	0.1348 W/(m·K)
Dielectric constant (25°C)	2.379
Relative vapor density (air = 1)	3.14
Vapor pressure (25°C)	0.0036 MPa
Solubility in water	0.047%, mass/mass
Flash point (CC)	4°C
Autoignition temperature	552°C
Explosive limits in air	1.4–7.4%, vol/vol
CAS registry number	108-88-3
Exposure limits	200 ppm, 8-hr TWA
Solubility parameter, δ	8.9
Hydrogen bond index, λ	3.8
Solvatochromic α	0.00
Solvatochromic β	0.11
Solvatochromic π^*	0.54

Notes: Aromatic solvent; sweet pungent odor; soluble in benzene, alcohols, organic halides, ethers; insoluble in water; highly flammable; toxic by ingestion, inhalation, and absorption through the skin; narcotic at high concentrations; incompatible with strong oxidants; decomposes under high heat to form (predominantly) dimethylbiphenyl.

Synonyms: toluol, methylbenzene, methylbenzol, phenylmethane.

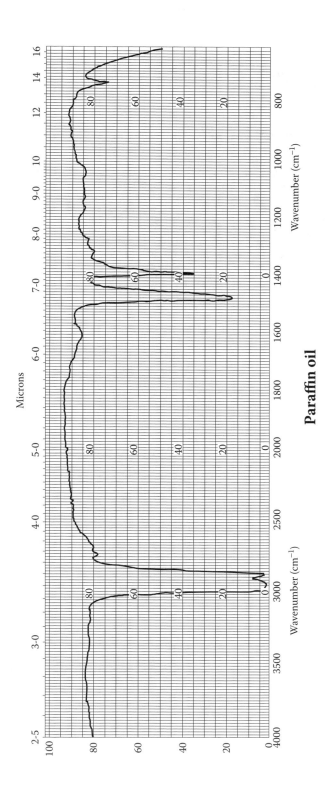

Paraffin oil

Figure 2.52

PARAFFIN OIL

Physical Properties

Relative molecular mass	Variable
Melting point	−20°C (approximate)
Normal boiling point	315°C (approximate)
Refractive index	
(20°C)	1.4720
(25°C)	1.4697
Specific gravity, 25°C/25°C	0.85
Solubility in water	Insoluble
Flash point (OC)	229°C
Explosive limits in air	0.6–6.5%, vol/vol
CAS registry number	8012-95-1
Exposure limits	50 ppm, 8-hr TWA

Notes: Viscous, odorless, moderately combustible liquid used for mull preparation; relatively low toxicity; soluble in benzene, chloroform, carbon disulfide, ethers; incompatible with oxidizing materials and amines.

Synonyms: mineral oil, adepsine oil, lignite oil, nujol.

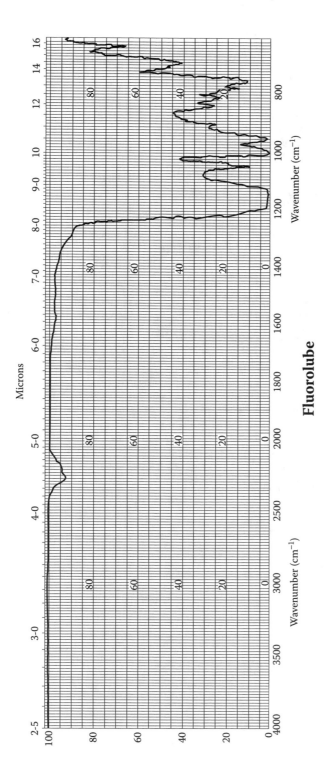

Fluorolube

Figure 2.53

FLUOROLUBE™, POLYTRIFLUOROCHLOROETHYLENE, $[-C_2ClF_3-]$

Physical Properties

Relative molecular mass (monomer)	116.47
Pour point*	−60 to 13°C
Melting point	−51 to 18°C
Acidity (pH)*	6.0–7.5
Density (38°C)*	1.865–1.955 g/mL
Viscosity (25°C)*	6–1400 mPa·s
Vapor pressure (93°C)	0.07–2.2 mm Hg
Flash point (OC)	Nonflammable
Autoignition temperature	Nonflammable
Explosive limits in air	Nonflammable
CAS registry number	9002-83-9
Exposure limits	Not established

Notes: There are six common grades or varieties of this oil, marketed under the name Fluorolube. The properties listed above that are marked with an asterisk depend upon the grade that is used. The primary physical differences between the grades are the viscosities and pour points.

The thermal stability of these materials is dependent on the wetted surfaces. Typical ranges of stability are between 150 and 325°C, but this varies with the wetted surface and residence time. Some metals can accelerate the decomposition into lower-molecular-mass, more volatile components. It is important to avoid the wetting of metals containing aluminum or magnesium, especially in situations in which high friction of galling is possible. Detonation of these fluids is possible under these conditions. Moreover, these fluids can react violently in the presence of sodium, potassium, amines, hydrazine, liquid fluorine, and liquid chlorine.

Because these fluids are essentially transparent from 1360 to 4000 cm^{-1} (except for the absorption at 2321.9 cm^{-1}), they can be used as mulling agents when the bands of paraffin oil obscure or interfere with sample absorptions.

POLYSTYRENE WAVENUMBER CALIBRATION

The wavelength (wavenumber) of typical infrared spectrophotometers is often calibrated or verified with a standard spectrum of polystyrene, obtained from a polymer film. The wavelength match is also used as a diagnostic measure of spectrometer performance. The spectrum on the following page is that of polystyrene film (film thickness = 50 μm), and the table provides the wavenumber readings assigned to the peaks on the spectrum.[1,2]

Wavenumber Readings Assigned to the Peaks on the Spectrum

1	—	3027.1	8	—	1583.1
2	—	2924.0	9	—	1181.4
3	—	2850.7	10	—	1154.3
4	—	1944.0	11	—	1069.1
5	—	1871.0	12	—	1028.0
6	—	1801.6	13	—	906.7
7	—	1601.4	14	—	698.9

Note: Film thickness = 50 μm.

REFERENCES

1. Bruno, T.J. and Svoronos, P.D.N., *CRC Handbook of Basic Tables for Chemical Analysis,* 2nd ed., CRC Press, Boca Raton, FL, 2003.
2. NIST, NIST Standard Reference Material 1921a; available on-line at http://srmors.nist.gov.

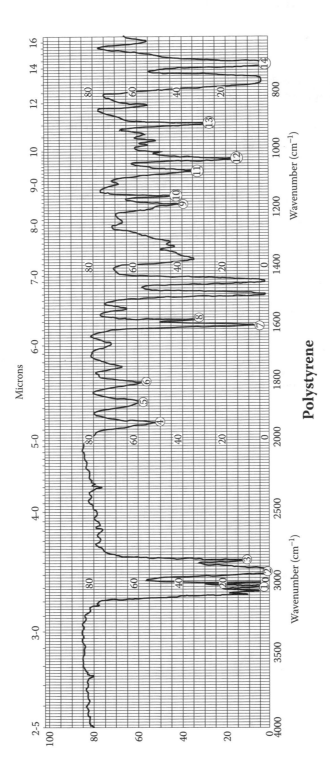

Polystyrene

Figure 2.54

WAVELENGTH-WAVENUMBER CONVERSION TABLE

The following table provides a conversion between wavelength and wavenumber units, for use in infrared spectrophotometry.

Wavelength-Wavenumber Conversion Table

Wavelength, μm	Wavenumber, cm⁻¹									
	0	1	2	3	4	5	6	7	8	9
2.0	5000	4975	4950	4926	4902	4878	4854	4831	4808	4785
2.1	4762	4739	4717	4695	4673	4651	4630	4608	4587	4566
2.2	4545	4525	4505	4484	4464	4444	4425	4405	4386	4367
2.3	4348	4329	4310	4292	4274	4255	4237	4219	4202	4184
2.4	4167	4149	4232	4115	4098	4082	4065	4049	4032	4016
2.5	4000	3984	3968	4953	3937	3922	3006	3891	3876	3861
2.6	3846	3831	3817	3802	3788	3774	3759	3745	3731	3717
2.7	3704	3690	3676	3663	3650	3636	3623	3610	3597	3584
2.8	3571	3559	3546	3534	3521	3509	3497	3484	3472	3460
2.9	3448	3436	3425	3413	3401	3390	3378	3367	3356	3344
3.0	3333	3322	3311	3300	3289	3279	3268	3257	3247	3236
3.1	3226	3215	3205	3195	3185	3175	3165	3155	3145	3135
3.2	3125	3115	3106	3096	3086	3077	3067	3058	3049	3040
3.3	3030	3021	3012	3003	2994	2985	2976	2967	2959	2950
3.4	2941	2933	2924	2915	2907	2899	2890	2882	2874	2865
3.5	2857	2849	2841	2833	2825	2817	2809	2801	2793	2786
3.6	2778	2770	2762	2755	2747	2740	2732	2725	2717	2710
3.7	2703	2695	2688	2681	2674	2667	2660	2653	2646	2639
3.8	2632	2625	2618	2611	2604	2597	2591	2584	2577	2571
3.9	2654	2558	2551	2545	2538	2532	2525	2519	2513	2506
4.0	2500	2494	2488	2481	2475	2469	2463	2457	2451	2445
4.1	2439	2433	2427	2421	2415	2410	2404	2398	2387	2387
4.2	2381	2375	2370	2364	2358	2353	2347	2342	2336	2331
4.3	2326	2320	2315	2309	2304	2299	2294	2288	2283	2278
4.4	2273	2268	2262	2257	2252	2247	2242	2237	2232	2227
4.5	2222	2217	2212	2208	2203	2198	2193	2188	2183	2179
4.6	2174	2169	2165	2160	2155	2151	2146	2141	2137	2132
4.7	2128	2123	2119	2114	2110	2105	2101	2096	2092	2088
4.8	2083	2079	2075	2070	2066	2062	2058	2053	2049	2045
4.9	2041	2037	2033	2028	2024	2020	2016	2012	2008	2004
5.0	2000	1996	1992	1988	1984	1980	1976	1972	1969	1965
5.1	1961	1957	1953	1949	1946	1942	1938	1934	1931	1927
5.2	1923	1919	1916	1912	1908	1905	1901	1898	1894	1890
5.3	1887	1883	1880	1876	1873	1869	1866	1862	1859	1855
5.4	1852	1848	1845	1842	1838	1835	1832	1828	1825	1821
5.5	1818	1815	1812	1808	1805	1802	1799	1795	1792	1788
5.6	1786	1783	1779	1776	1773	1770	1767	1764	1761	1757
5.7	1754	1751	1748	1745	1742	1739	1736	1733	1730	1727
5.8	1724	1721	1718	1715	1712	1709	1706	1704	1701	1698
5.9	1695	1692	1689	1686	1684	1681	1678	1675	1672	1669
6.0	1667	1664	1661	1668	1656	1653	1650	1647	1645	1642
6.1	1639	1637	1634	1631	1629	1626	1623	1621	1618	1616
6.2	1613	1610	1608	1605	1603	1600	1597	1595	1592	1590
6.3	1587	1585	1582	1580	1577	1575	1572	1570	1567	1565
6.4	1563	1560	1558	1555	1553	1550	1548	1546	1543	1541
6.5	1538	1536	1534	1531	1529	1527	1524	1522	1520	1517
6.6	1515	1513	1511	1508	1506	1504	1502	1499	1497	1495
6.7	1493	1490	1488	1486	1484	1481	1479	1477	1475	1473
6.8	1471	1468	1466	1464	1462	1460	1458	1456	1453	1451
6.9	1449	1447	1445	1443	1441	1439	1437	1435	1433	1431
7.0	1429	1427	1425	1422	1420	1418	1416	1414	1412	1410
7.1	1408	1406	1404	1403	1401	1399	1397	1395	1393	1391
7.2	1389	1387	1385	1383	1381	1379	1377	1376	1374	1372
7.3	1370	1368	1366	1364	1362	1361	1359	1357	1355	1353
7.4	1351	1350	1348	1346	1344	1342	1340	1339	1337	1335
7.5	1333	1332	1330	1328	1326	1325	1323	1321	1319	1318
7.6	1316	1314	1312	1311	1309	1307	1305	1304	1302	1300
7.7	1299	1297	1295	1294	1292	1290	1289	1287	1285	1284
7.8	1282	1280	1279	1277	1276	1274	1272	1271	1269	1267
7.9	1266	1264	1263	1261	1259	1258	1256	1255	1253	1252

(continued)

Wavelength-Wavenumber Conversion Table (Continued)

Wavelength, μm	Wavenumber, cm⁻¹									
	0	1	2	3	4	5	6	7	8	9
8.0	1250	1248	1247	1245	1244	1242	1241	1239	1238	1236
8.1	1235	1233	1232	1230	1229	1227	1225	1224	1222	1221
8.2	1220	1218	1217	1215	1214	1212	1211	1209	1208	1206
8.3	1205	1203	1202	1200	1199	1198	1196	1195	1193	1192
8.4	1190	1189	1188	1186	1185	1183	1182	1181	1179	1178
8.5	1176	1175	1174	1172	1171	1170	1168	1167	1166	1164
8.6	1163	1161	1160	1159	1157	1156	1155	1153	1152	1151
8.7	1149	1148	1147	1145	1144	1143	1142	1140	1139	1138
8.8	1136	1135	1134	1133	1131	1130	1129	1127	1126	1125
8.9	1124	1122	1121	1120	1119	1117	1116	1115	1114	1112
9.0	1111	1110	1109	1107	1106	1105	1104	1103	1101	1100
9.1	1099	1098	1096	1095	1094	1093	1092	1091	1089	1088
9.2	1087	1086	1085	1083	1082	1081	1080	1079	1078	1076
9.3	1075	1074	1073	1072	1071	1070	1068	1067	1066	1065
9.4	1064	1063	1062	1060	1059	1058	1057	1056	1055	1054
9.5	1053	1052	1050	1049	1048	1047	1046	1045	1044	1043
9.6	1042	1041	1040	1038	1037	1036	1035	1034	1033	1032
9.7	1031	1030	1029	1028	1027	1026	1025	1024	1022	1021
9.8	1020	1019	1018	1017	1016	1015	1014	1013	1012	1011
9.9	1010	1009	1008	1007	1006	1005	1004	1003	1002	1001
10.0	1000	999	998	997	996	995	994	993	992	991
10.1	990	989	988	987	986	985	984	983	982	981
10.2	980	979	978	978	977	976	975	974	973	972
10.3	971	970	969	968	967	966	965	964	963	962
10.4	962	961	960	959	958	957	956	955	954	953
10.5	952	951	951	950	949	948	947	946	945	944
10.6	943	943	942	941	940	939	938	937	936	935
10.7	935	934	933	932	931	930	929	929	928	927
10.8	926	925	924	923	923	922	921	920	919	918
10.9	917	917	916	915	914	913	912	912	911	910
11.0	909	908	907	907	906	905	904	903	903	902
11.1	901	900	899	898	898	897	896	895	894	894
11.2	893	892	891	890	890	889	888	887	887	886
11.3	885	884	883	883	882	881	880	880	879	878
11.4	877	876	876	875	874	873	873	872	871	870
11.5	870	869	868	867	867	866	865	864	864	863
11.6	862	861	861	860	859	858	858	857	856	855
11.7	855	854	853	853	852	851	850	850	849	848
11.8	847	847	846	845	845	844	843	842	842	841
11.9	840	840	839	838	838	837	836	835	835	834
12.0	833	833	832	831	831	830	829	829	828	827
12.1	826	826	825	824	824	823	822	822	821	820
12.2	820	819	818	818	817	816	816	815	814	814
12.3	813	812	812	811	810	810	809	808	808	807
12.4	806	806	805	805	804	803	803	802	801	801
12.5	800	799	799	798	797	797	796	796	795	794
12.6	794	793	792	792	791	791	790	789	789	788
12.7	787	787	786	786	785	784	784	783	782	782
12.8	781	781	780	779	779	778	778	777	776	776
12.9	775	775	774	773	773	772	772	771	770	770
13.0	769	769	768	767	767	766	766	765	765	764
13.1	763	763	762	762	761	760	760	759	759	758
13.2	758	757	756	756	755	755	754	754	753	752
13.3	752	751	751	750	750	749	749	748	747	747
13.4	746	746	745	745	744	743	743	742	742	741
13.5	741	740	740	739	739	738	737	737	736	736
13.6	735	735	734	734	733	733	732	732	731	730
13.7	730	729	729	728	728	727	727	726	726	725
13.8	725	724	724	723	723	722	722	721	720	720
13.9	719	719	718	718	717	717	716	716	715	715

(continued)

Wavelength-Wavenumber Conversion Table (Continued)

Wavelength, μm	Wavenumber, cm⁻¹									
	0	1	2	3	4	5	6	7	8	9
14.0	714	714	713	713	712	712	711	711	710	710
14.1	709	709	708	708	707	707	706	706	705	705
14.2	704	704	703	703	702	702	702	701	701	700
14.3	699	699	698	698	697	697	696	696	695	695
14.4	694	694	693	693	693	692	692	691	691	690
14.5	690	689	689	688	688	687	687	686	686	685
14.6	685	684	684	684	683	683	682	682	681	681
14.7	680	680	679	679	678	678	678	677	677	676
14.8	676	675	675	674	674	673	673	672	672	672
14.9	671	671	670	670	669	669	668	668	668	667

DIAGNOSTIC SPECTRA

The interpretation of infrared spectra is often complicated by the presence of spurious absorbances or by instrumental upset conditions that must be recognized. In these cases, it is often helpful to refer to the spectra of common compounds that may be the cause of such difficulties. The following spectra present such diagnostic tools.[1] Carbon dioxide, as an atmospheric constituent, is often present as an unwanted contaminant. Water is also an atmospheric constituent, and is also present in many chemical processes. It can also react with certain species such as amines.

REFERENCE

1. NIST, NIST Chemistry Web Book, NIST Standard Reference Database, No. 69, March 2003 http://webbook.nist.gov/chemistry.

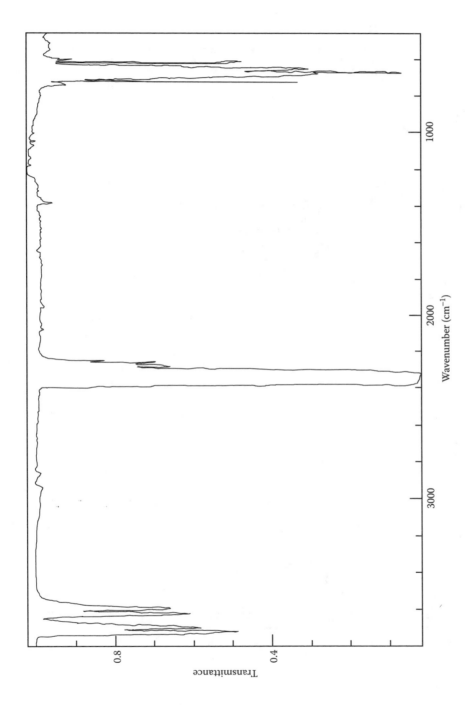

Figure 2.55 Infrared spectrum of carbon dioxide.

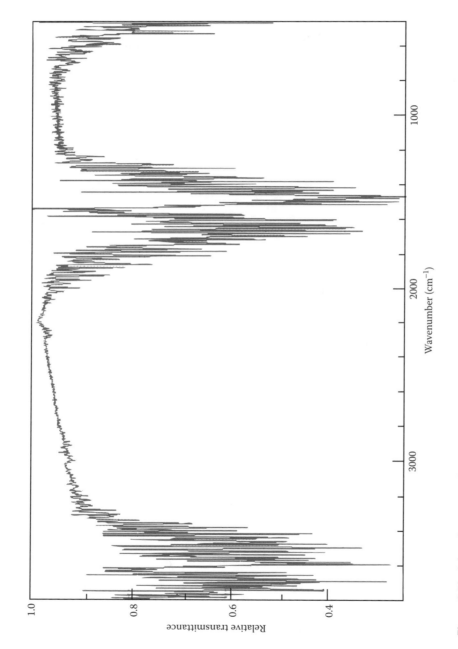

Figure 2.56   Infrared spectrum of water.

CHAPTER 3

Nuclear Magnetic Resonance Spectroscopy

CONTENTS

CHEMICAL-SHIFT RANGES OF SOME NUCLEI

The following table gives an approximate chemical-shift range (in ppm) for some of the most commonly studied nuclei. The range is established by the shifts recorded for the most common compounds.[1-12] The natural abundance of the nucleus and the chemical-shift range are two of the most important factors taken into consideration when choosing a nucleus for chemical analysis.

REFERENCES

1. Yoder, C.H. and Schaeffer, C.D., Jr., *Introduction to Multinuclear NMR,* Benjamin/Cummings Publishing, Menlo Park, CA, 1987.
2. Silverstein, R.M., Bassler, G.C., and Morrill, T.C., *Spectrometric Identification of Organic Compounds,* 5th ed., John Wiley and Sons, New York, 1991.
3. Harris, R.U. and Mann, B.E., *NMR and the Periodic Table,* Academic Press, London, 1978.
4. Silverstein, R.M. and Webster F.X., *Spectrometric Identification of Organic Compounds,* 6th ed., John Wiley and Sons, New York, 1998.
5. Gunther, H., *NMR Spectroscopy: Basic Principles, Concepts, and Applications in Chemistry,* John Wiley and Sons, New York, 2003.
6. Kitamaru, R., *Nuclear Magnetic Resonance: Principles and Theory,* Elsevier Science, New York, 1990.
7. Lambert, J.B., Holland, L.N., and Mazzola, E.P., *Nuclear Magnetic Resonance Spectroscopy: Introduction to Principles, Applications and Experimental Methods,* Prentice Hall, Englewood Cliffs, NJ, 2003.
8. Bovey, F.A. and Mirau, P.A., *Nuclear Magnetic Resonance Spectroscopy,* 2nd ed., Academic Press, New York, 1988.
9. Harris, R.K. and Mann, B.E., *NMR and the Periodic Table,* Academic Press, London, 1978.
10. Hore, P.J. and Hore, P.J., *Nuclear Magnetic Resonance,* Oxford University Press, Oxford, 1995.
11. Nelson, J.H., *Nuclear Magnetic Resonance Spectroscopy,* 2nd ed., John Wiley and Sons, New York, 2003.
12. Bruno, T.J. and Svoronos, P.D.N., *Handbook of Basic Tables for Chemical Analysis,* 2nd ed., CRC Press, Boca Raton, FL, 2003.

Chemical-Shift Range for Commonly Studied Nuclei

Nucleus	Chemical-Shift Range, ppm	Natural Abundance, %	Nucleus	Chemical-Shift Range, ppm	Natural Abundance, %
^{1}H	15	99.9844	^{29}Si	400	4.70
^{7}Li	10	92.57	^{31}P	700	100
^{11}B	200	81.17	^{33}S	600	0.74
^{13}C	250	1.108	^{35}Cl	820	75.4
^{15}N	930	0.365	^{39}K	60	93.08
^{17}O	700	0.037	^{59}Co	14,000	100
^{19}F	800	100	^{119}Sn	2,000	8.68
^{23}Na	15	100	^{133}Cs	150	100
^{27}Al	270	100	^{207}Pb	10,000	21.11

REFERENCE STANDARDS FOR SELECTED NUCLEI

The following table lists the most important and commonly used chemical standards that are employed when NMR spectra of various nuclei are being measured. The standards should be inert, soluble in a variety of solvents, and preferably should produce one singlet peak that appears close to the lowest-frequency region of the chemical-shift range. When NMR data are provided or interpreted, it is always necessary to specify the reference standard employed.[1–5] The choice of the reference standard is usually made to fit the polarity of both the solute and solvent so that they will yield a homogeneous solution.

REFERENCES

1. Yoder, C.H. and Schaeffer, C.D., Jr., *Introduction to Multinuclear NMR*, Benjamin/Cummings Publishing, Menlo Park, CA, 1987.
2. Grim, S.O. and Yankowsky, A.W., On the phosphorus-31 chemical shifts of substituted triarylphosphines, *Phosphorus and Sulfur*, 3, 191, 1977.
3. Gunther, H., *NMR Spectroscopy: Basic Principles, Concepts and Applications in Chemistry*, John Wiley and Sons, New York, 2003.
4. Abraham, R.J., Fisher, J., and Loftus, P., *Introduction to NMR Spectroscopy*, John Wiley and Sons, New York, 1988.
5. Bruno, T.J. and Svoronos, P.D.N., *Handbook of Basic Tables for Chemical Analysis*, 2nd ed., CRC Press, Boca Raton, FL, 2003.

Chemical Standards Used When Measuring NMR Spectra

Nucleus	Name	Formula
^1H	Tetramethylsilane (TMS)	$(CH_3)_4Si$
	3-(Trimethylsilyl)-1-propanesulfonic acid, sodium salt (DSS)[a]	$(CH_3)_3Si(CH_2)_3SO_3Na$
	3-(Trimethylsilyl)-propanoic acid, d_4, sodium salt (TSP)	$(CH_3)_3Si(CD_2)_3CO_2Na$
^2H	Deuterated chloroform (chloroform-d)	$CDCl_3$
^{11}B	Boric acid	H_3BO_3
	Boron trifluoride etherate	$(C_2H_5)_2O \cdot BF_3$
	Boron trichloride	BCl_3
^{13}C	Tetramethylsilane (TMS)	$(CH_3)_4Si$
^{15}N	Ammonium nitrate	NH_4NO_3
	Ammonia	NH_3
	Nitromethane	CH_3NO_2
	Nitric acid	HNO_3
	Tetramethylammonium chloride	$(CH_3)_4NCl$
^{17}O	Water	H_2O
^{19}F	Trichlorofluoromethane (Freon 11, R-11)	CCl_3F
	Hexafluorobenzene	C_6F_6
^{31}P	Trimethylphosphite (methyl phosphite)	$(CH_3O)_3P$
	Phosphoric acid (85%)	H_3PO_4
^{35}Cl	Sodium chloride	$NaCl$
^{59}Co	Cobalt (III) hexacyanide anion	$K_3Co(CN)_6$
^{119}Sn	Tetramethyltin	$(CH_3)_4Sn$
^{195}Pt	Platinum (IV) hexacyanide	$K_2Pt(CN)_6$
	Dihydrogen platinum (IV) hexachloride	H_2PtCl_6
^{183}W	Sodium tungstate (external)	Na_2WO_4

[a] For aqueous solutions (known also as "water-soluble TMS" or 2,2-dimethyl-2-silapentane-5-sulfonate).

¹H AND ¹³C CHEMICAL SHIFTS OF USEFUL NMR SOLVENTS

The following table lists the expected $^1H(\delta_H)$ and $^{13}C(\delta_C)$ chemical shifts for various useful liquid NMR solvents in parts per million (ppm). The table also includes the liquid temperature range (°C) and dielectric constants of these solvents. Slight changes may occur with changes in concentration, especially in acidic hydrogens, as indicated in the following table.[1-4]

REFERENCES

1. Silverstein, R.M., Bassler, G.C., and Morrill, T.C., *Spectrometric Identification of Organic Compounds,* 5th ed., John Wiley and Sons, New York, 1991.
2. Rahman, A.-U., *Nuclear Magnetic Resonance: Basic Principles,* Springer-Verlag, New York, 1986.
3. Abraham, R.J., Fisher, J., and Loftus, P., *Introduction of NMR Spectroscopy,* John Wiley and Sons, Chichester, U.K., 1988.
4. Bruno, T.J. and Svoronos, P.D.N., *Handbook of Basic Tables for Chemical Analysis,* 2nd ed., CRC Press, Boca Raton, FL, 2003.

Chemical Shifts for Useful Liquid NMR Solvents

Solvent	Formula	Liquid Temperature Range, °C	Dielectric Constant (ε)	δ_H, ppm	δ_C, ppm
Acetone-d₆	$(CD_3)_2CO$	−95 to 56	20.7	2.17	29.2, 204.1
Acetonitrile-d₃	CD_3CN	−44 to 82	37.5	2.00	1.3, 117.7
Benzene-d₆	C_6D_6	6 to 80	2.284	7.27	128.4
Carbon disulfide	CS_2	−112 to 46	2.641	—	192.3
Carbon tetrachloride	CCl_4	−23 to 77	2.238	—	96.0
Chloroform-d₃	$CDCl_3$	−64 to 61	4.806	7.25	76.9
Cyclohexane-d₁₂	C_6D_{12}	6 to 81	2.023	1.43	27.5
Dichloromethane-d₂	CD_2Cl_2	−95 to 40	9.08	5.33	53.6
Difluorobromochloromethane	CF_2BCl	−140 to −25	—	—	109.2
Dimethylformamide-d₇	$DCON(CD_3)_2$	−60 to 153	36.7	2.9, 3.0, 8.0	31, 36, 132.4
Dimethylsulfoxide-d₆	$(CD_3)_2SO$	19 to 189	46.7	2.62	39.6
1,4-Dioxane-d₈	$C_4D_8O_2$	12 to 101	2.209	3.7	67.4
HMPA	—	7 to 233	30.0	2.60	36.8
Methanol-d₄	CD_3OD	−98 to 65	32.63	3.4, 4.8ᵃ	49.3
Nitrobenzene	$C_6D_5NO_2$	6 to 211	34.8	8.2, 7.6, 7.5	149, 134, 129, 124
Nitromethane-d₃	CD_3NO_2	−29 to 101	35.87	4.33	57.3
Pyridine-d₅	C_5D_5N	−42 to 115	123	7.0, 7.6, 8.6	124, 136, 150
1,1,2,2-Tetrachloroethane-d₂	CD_2ClCD_2Cl	−44 to 146	8.2	5.94	75.5
Tetrahydrofuran-d₈	C_4D_8O	−108 to 66	7.54	1.9, 3.8	25.8, 67.9
1,2,4-Trichlorobenzene	$C_6D_3Cl_3$	17 to 214	3.9	7.1, 7.3, 7.4	133.3, 132.8, 130.7, 130.0, 127.6
Trichlorofluoromethane	$CFCl_3$	−111 to 24	2.3	—	117.6
Vinyl chloride-d₃	$CD_2=CDCl$	−154 to −13	—	5.4, 5.5, 6.3	126, 117
Trifluoroacetic acid, d	CF_3COOD	−15 to 72	8.6	11.3 ᵃ	114.5, 116.5
Water-d₂	D_2O	0 to 100	78.5	4.7 ᵃ	—

ᵃ Variable with concentration.

PROTON NMR ABSORPTIONS OF MAJOR FUNCTIONAL GROUPS

The following correlation tables provide the regions of nuclear magnetic resonance absorptions of major chemical families. These absorptions are reported in the dimensionless units of parts per million (ppm) versus the standard compound tetramethylsilane (TMS, $(CH_3)_4Si$), which is recorded as 0.0 ppm.

The use of this unit of measure makes the chemical shifts independent of the applied magnetic field strength or the radio frequency. For most proton NMR spectra, the protons in TMS are more shielded than almost all other protons. The chemical shift in this dimensionless unit system is then defined by

$$\delta = \frac{\nu_s - \nu_r}{\nu_r} \times 10^6$$

where ν_s and ν_r are the absorption frequencies of the sample proton and the reference (TMS) protons (12, magnetically equivalent), respectively. In these tables, the proton(s) whose proton NMR shifts are cited are indicated by an underscore. For more detail concerning these conventions, the reader is referred to the general references below.[1-12] Reference 13 has a compilation of references for the various nuclei.

Due to the large amount of data, the whole ^1H NMR region is divided into smaller sections of 1.0 to 1.2 ppm range each. This will allow the user to look into a specified chemical shift and determine all possibilities for the unknown whose structure is being analyzed. The symbol Ar refers to an aryl group.

REFERENCES

1. Silverstein, R.M. and Webster, F.X., *Spectrometric Identification of Organic Compounds,* 6th ed., Wiley, New York, 1998.
2. Rahman, A.-U., *Nuclear Magnetic Resonance,* Springer Verlag, New York, 1986.
3. Gordon, A.J. and Ford, R.A., *The Chemist's Companion,* Wiley Interscience, New York, 1971.
4. Becker, E.D., *High Resolution NMR, Theory and Chemical Applications,* 2nd ed., Academic Press, New York, 1980.
5. Gunther, H., *NMR Spectroscopy: Basic Principles, Concepts and Applications in Chemistry,* Wiley, New York, 2003.
6. Kitamaru, R., *Nuclear Magnetic Resonance: Principles and Theory,* Elsevier Science, New York, 1990.
7. Lambert, J.B., Holland, L.N., and Mazzola, E.P., *Nuclear Magnetic Resonance Spectroscopy: Introduction to Principles, Applications, and Experimental Methods,* Prentice Hall, Englewood Cliffs, NJ, 2003.
8. Bovey, F.A. and Mirau, P.A., *Nuclear Magnetic Resonance Spectroscopy,* 2nd ed., Academic Press, New York, 1988.
9. Hore, P.J. and Hore, P.J., *Nuclear Magnetic Resonance,* Oxford University Press, Oxford, 1995.
10. Bruno, T.J. and Svoronos, P.D.N., *Handbook of Basic Tables for Chemical Analysis,* 2nd ed., CRC Press, Boca Raton, FL, 2003.
11. Nelson, J.H., *Nuclear Magnetic Resonance Spectroscopy,* 2nd ed., Wiley, New York, 2003.
12. Abraham, R.J., Fisher, J., and Loftus, P., *Introduction to NMR Spectroscopy,* Wiley, New York, 1988.
13. University of Wisconsin, NMR Bibliography; available on-line at http://www.chem.wisc.edu/areas/reich/Handouts/nmr/NMR-Biblio.htm.

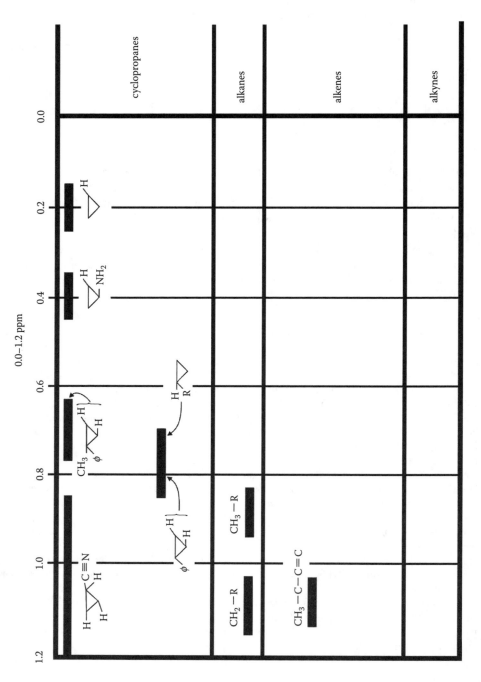

Figure 3.1

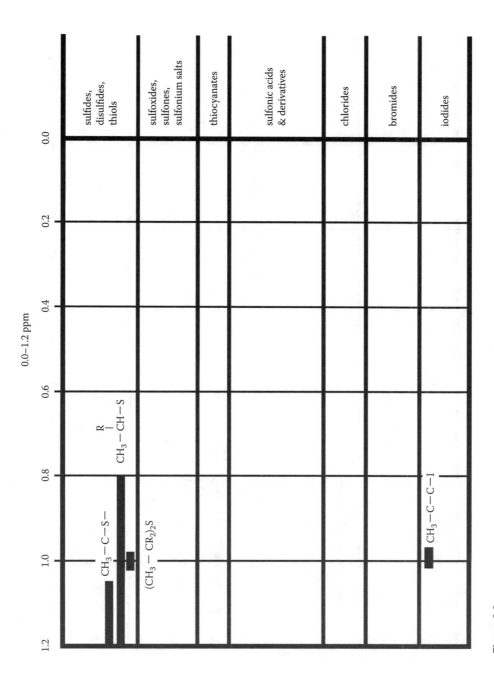

Figure 3.2

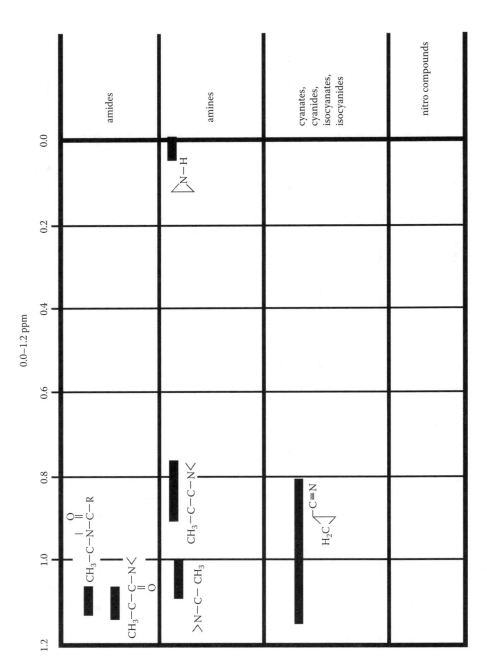

Figure 3.3

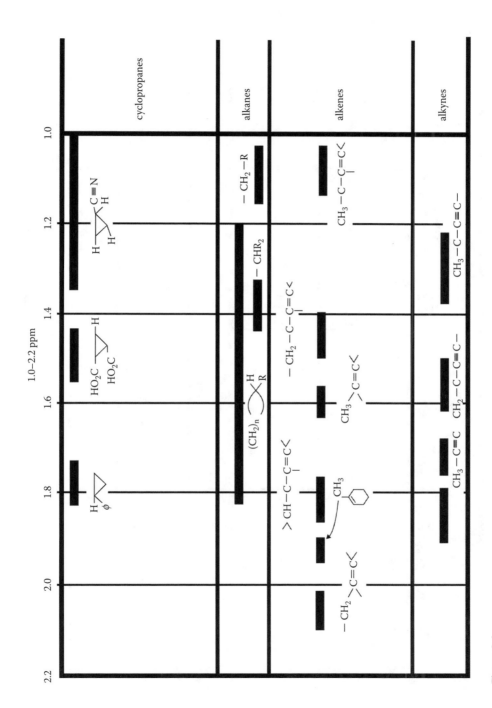

Figure 3.4

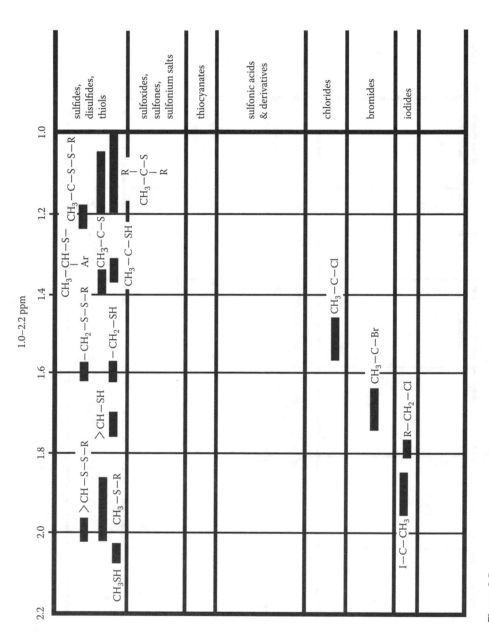

Figure 3.5

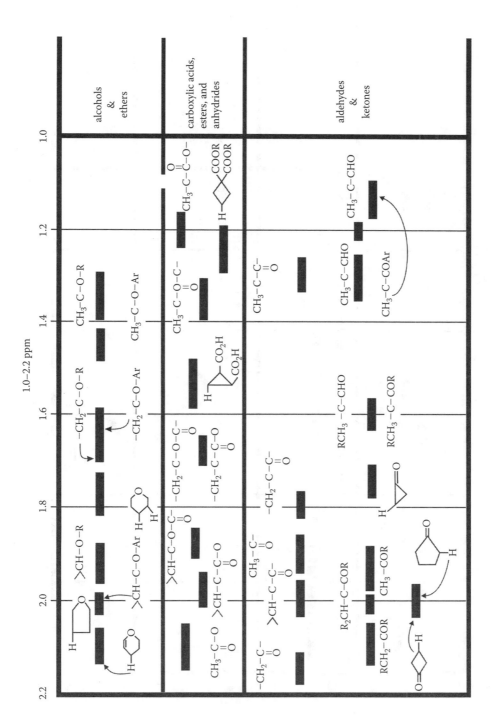

Figure 3.6

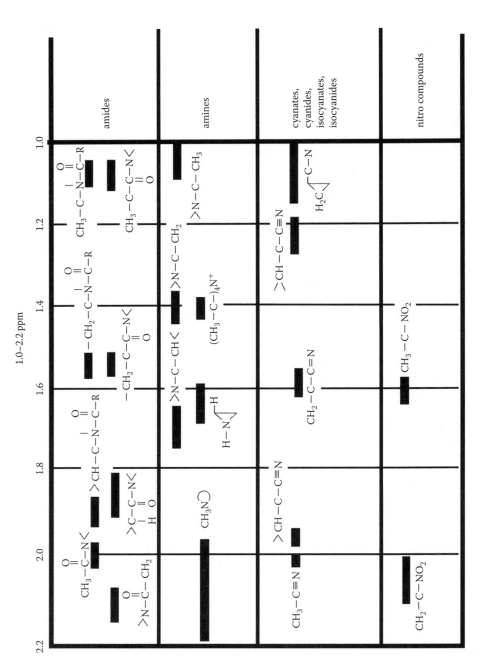

Figure 3.7

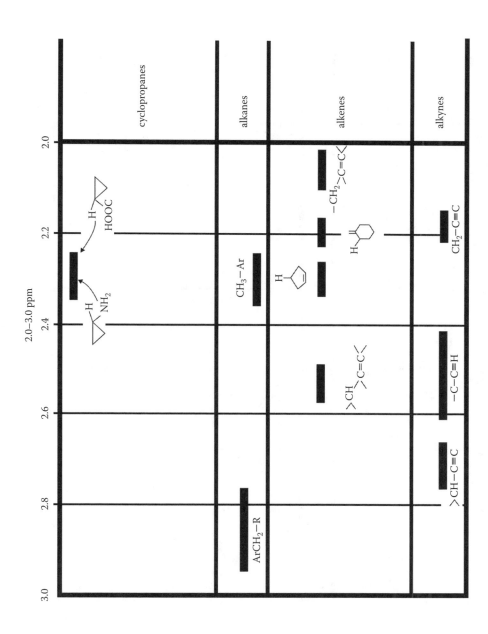

Figure 3.8

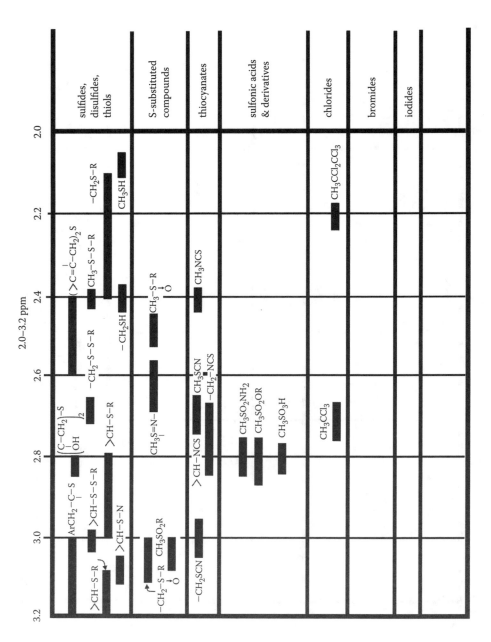

Figure 3.9

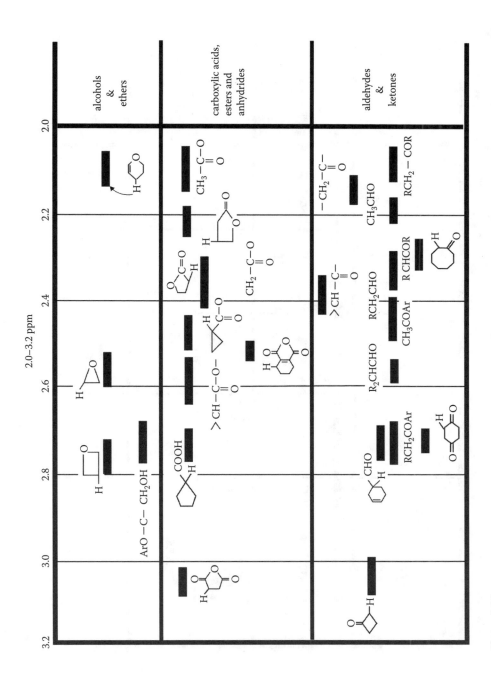

Figure 3.10

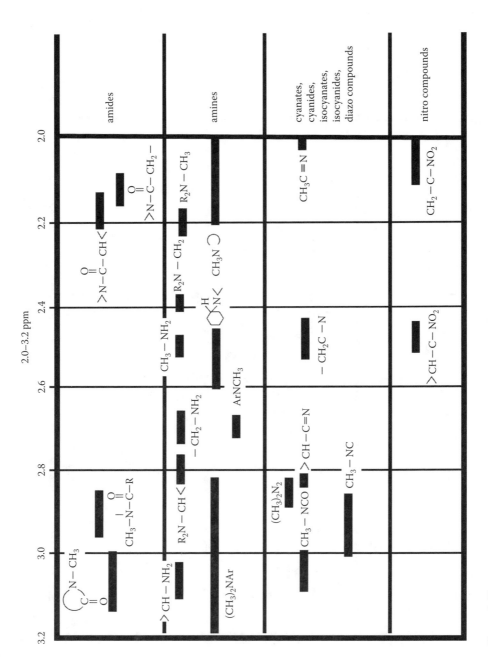

Figure 3.11

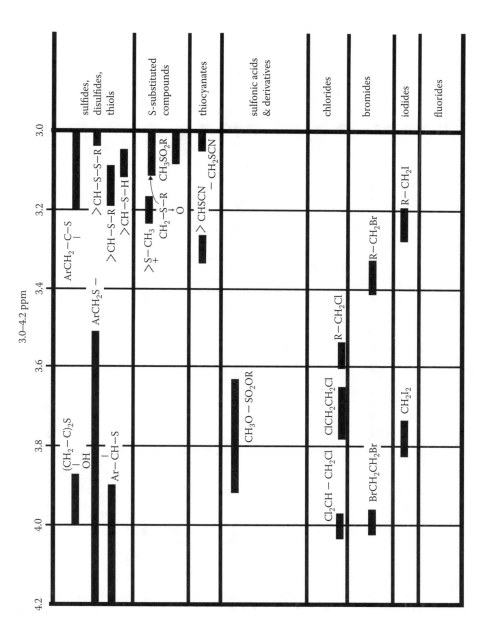

Figure 3.12

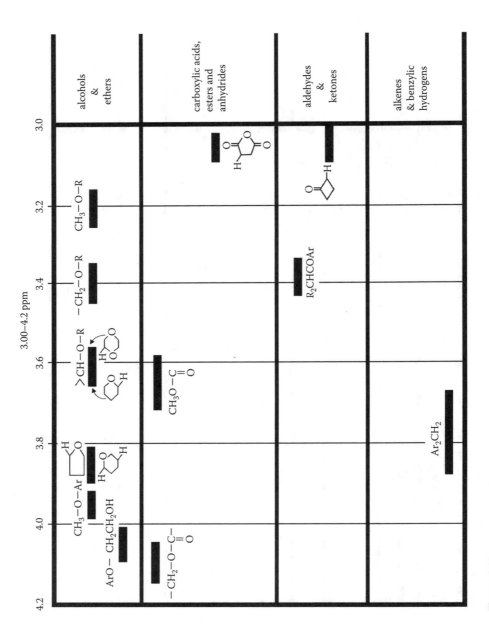

Figure 3.13

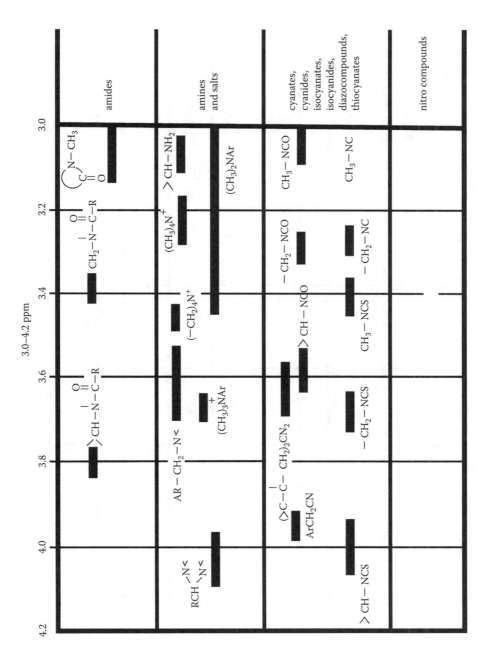

Figure 3.14

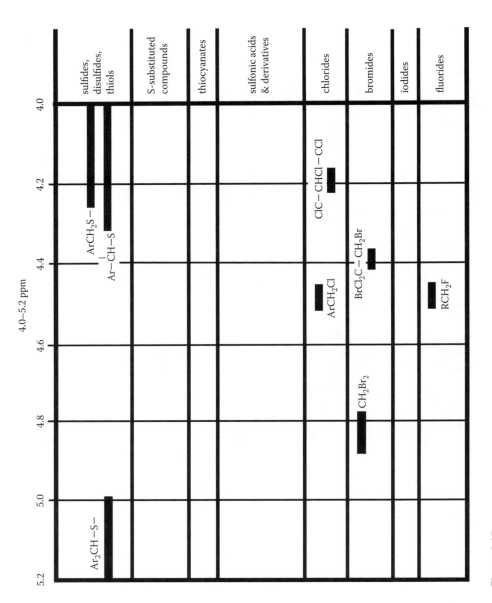

Figure 3.15

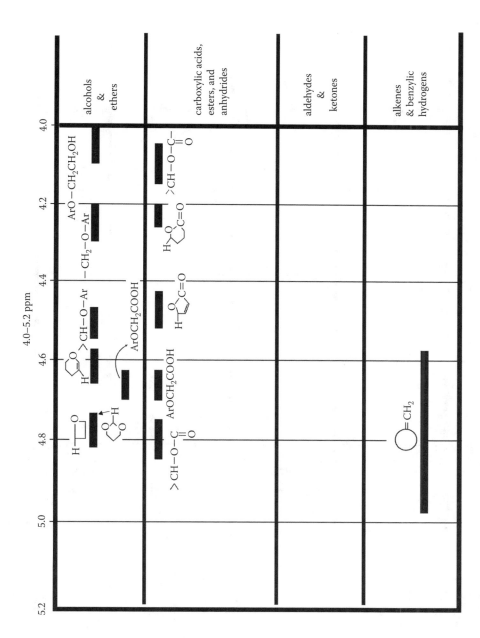

Figure 3.16

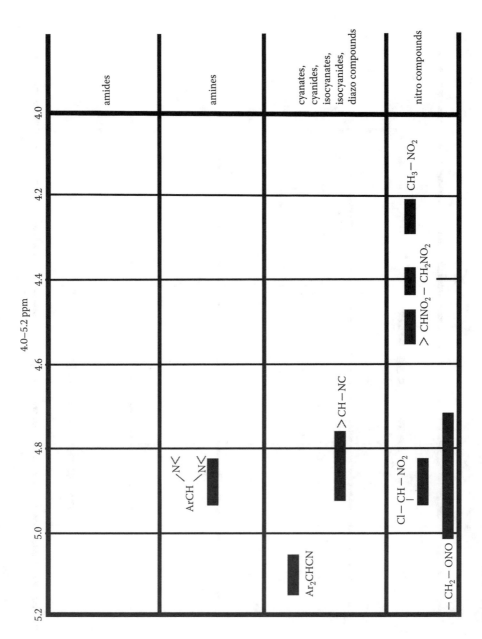

Figure 3.17

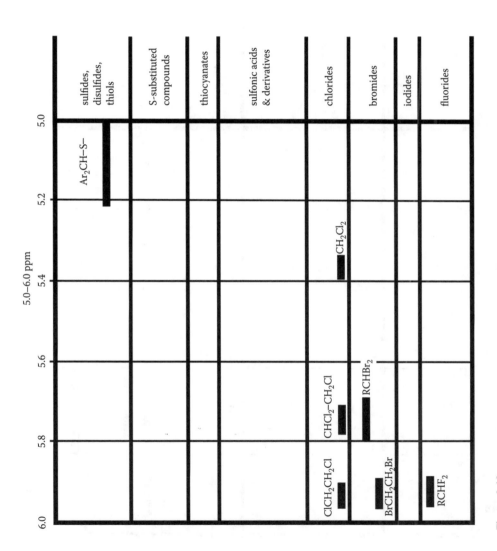

Figure 3.18

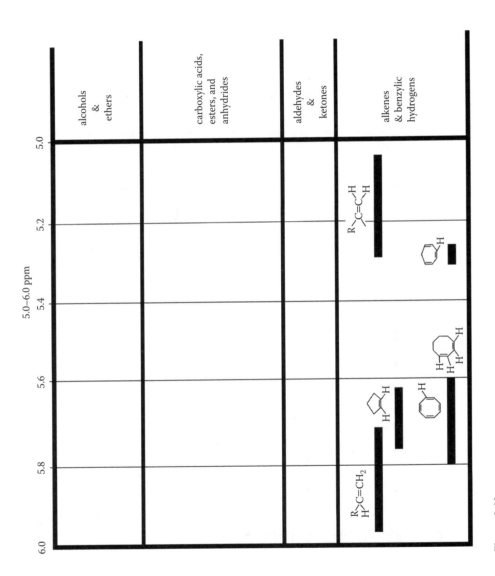

Figure 3.19

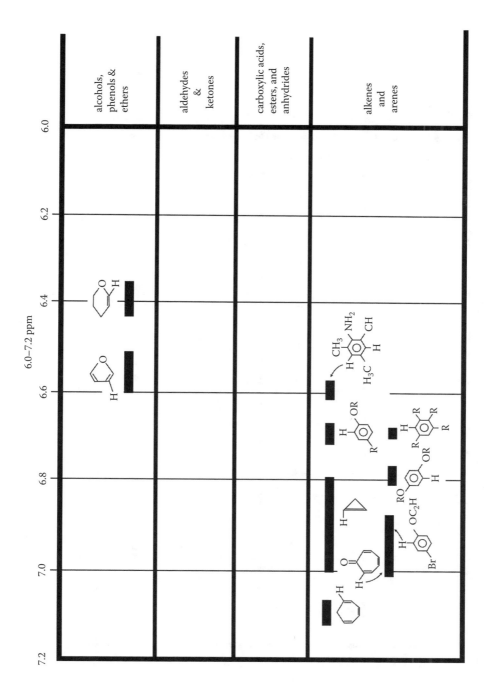

Figure 3.20

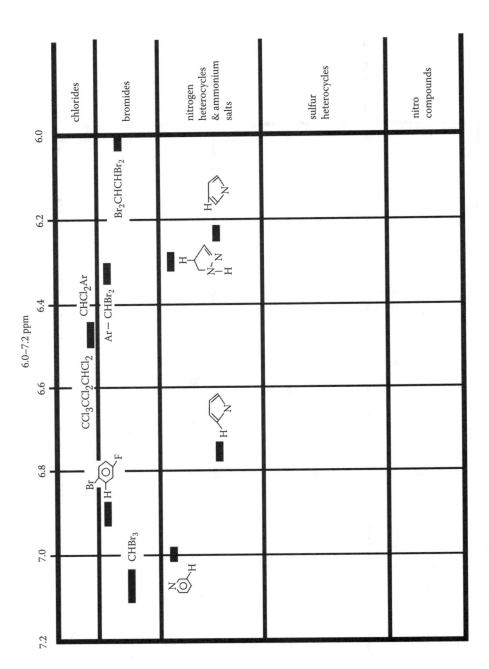

Figure 3.21

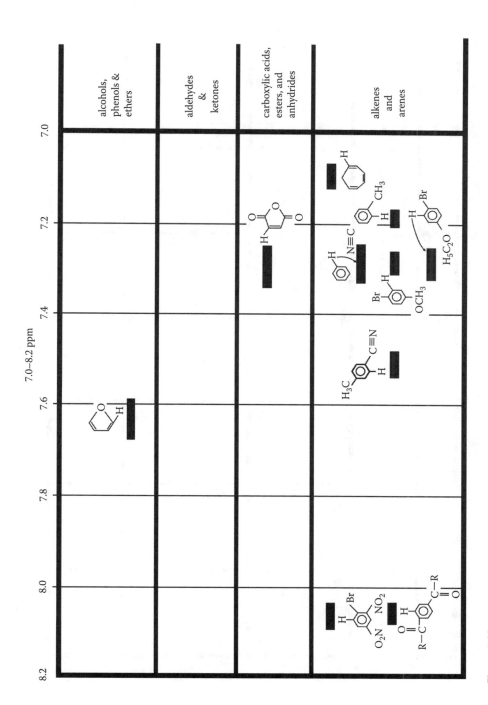

Figure 3.22

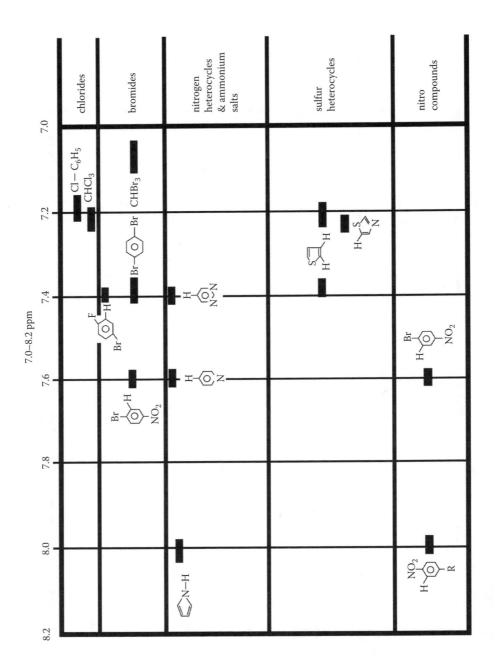

Figure 3.23

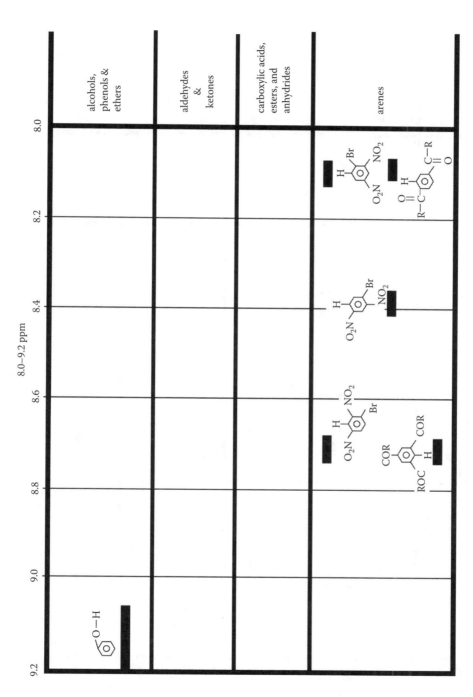

Figure 3.24

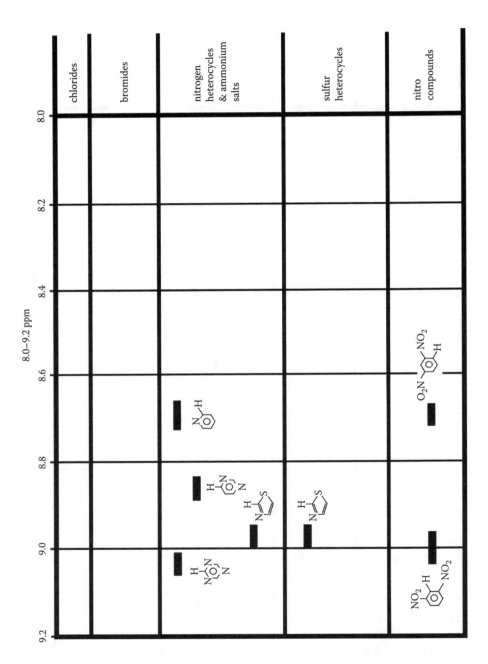

Figure 3.25

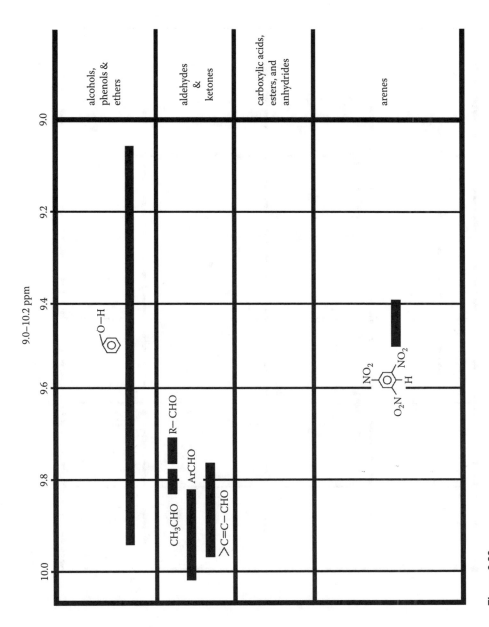

Figure 3.26

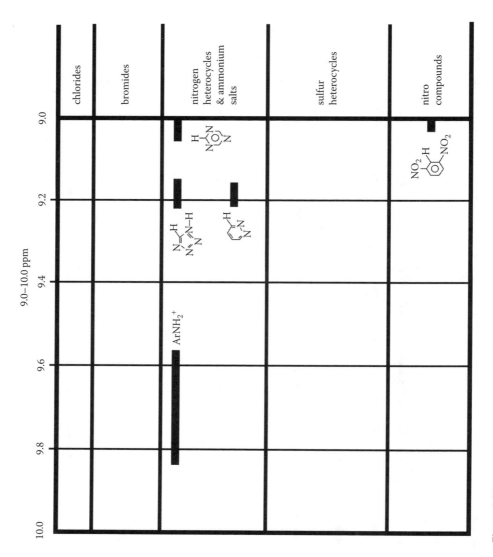

Figure 3.27

SOME USEFUL ¹H COUPLING CONSTANTS

This chart[1] provides the values of some useful proton NMR coupling constants (in hertz). The data are adapted with permission from the work of Dr. C.F. Hammer, professor emeritus, Chemistry Department, Georgetown University, Washington, DC 20057. The single numbers indicate a typical average, while in some cases, the range is provided.

REFERENCE

1. Bruno, T.J. and Svoronos, P.D.N., *Handbook of Basic Tables for Chemical Analysis,* 2nd ed., CRC Press, Boca Raton, FL, 2003.

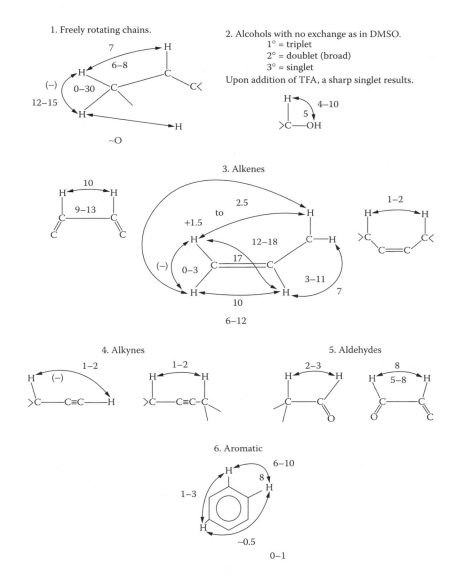

Figure 3.28

ADDITIVITY RULES IN ^{13}C-NMR CORRELATION TABLES

The wide chemical-shift range ($\approx$250 ppm) of ^{13}C NMR is responsible for the considerable change of a chemical shift noted when a slight inductive, mesomeric, or hybridization change occurs on a neighboring atom. Following the various empirical correlations in ^{1}H NMR,[1-7] D.W. Brown[8] has developed a short set of ^{13}C-NMR correlation tables. This section covers a part of those as adopted by Yoder and Schaeffer[9] and Clerk et al.[10] The reader is advised to refer to Brown,[8] and should the need for some specific data on more complicated structures arise, additional sources are provided.[11-18]

REFERENCES

1. Shoolery, J.N., *Varian Associates Technical Information Bulletin,* 2(3), 1959.
2. Bell, H.M., Bowles, D.B., and Senese, F., Additive NMR chemical shift parameters for deshielded methine protons, *Org. Magn. Res.,* 16, 285, 1981.
3. Matter, U.E., Pascual, C., Pretsch, E., Pross, A., Simon, W., and Sternhell, S., Estimation of the chemical shifts of olefinic protons using additive increments, II: Compilation of additive increments for 43 functional groups, *Tetrahedron,* 25, 691, 1969.
4. Matter, U.E., Pascual, C., Pretsch, E., Pross, A., Simon, W., and Sternhell, S., Estimation of the chemical shifts of olefinic protons using additive increments, III: Examples of utility in N.M.R. studies and the identification of some structural features responsible for deviations from additivity, *Tetrahedron,* 25, 2023, 1969.
5. Jeffreys, J.A.D., A rapid method for estimating NMR shifts for protons attached to carbon, *J. Chem. Educ.,* 56, 806, 1979.
6. Mikolajczyk, M., Grzeijszczak, S., and Zatorski, A., Organosulfur compounds, IX: NMR and structural assignments in α,β-unsaturated sulphoxides using additive increments method, *Tetrahedron,* 32, 969, 1976.
7. Friedrich, E.C. and Runkle, K.G., Empirical NMR chemical shift correlations for methyl and methylene protons, *J. Chem. Educ.,* 61, 830, 1984.
8. Brown, D.W., A short set of ^{13}C-NMR correlation tables, *J. Chem. Educ.,* 62, 209, 1985.
9. Yoder, C.H. and Schaeffer, C.D., Jr., *Introduction to Multinuclear NMR,* Benjamin/Cummings Publishing, Menlo Park, CA, 1987.
10. Clerk, J.T., Pretsch, E., and Seibl, J., *Structural Analysis of Organic Compounds by Combined Application of Spectroscopic Methods,* Elsevier, Amsterdam, 1981.
11. Silverstein, R.M. and Webster F.X., *Spectrometric Identification of Organic Compounds,* 6th ed., John Wiley and Sons, New York, 1998.
12. Gunther, H., *NMR Spectroscopy: Basic Principles, Concepts and Applications in Chemistry,* Wiley, New York, 2003.
13. Kitamaru, R., *Nuclear Magnetic Resonance: Principles and Theory,* Elsevier Science, New York, 1990.
14. Lambert, J.B., Holland, L.N., and Mazzola, E.P., *Nuclear Magnetic Resonance Spectroscopy: Introduction to Principles, Applications and Experimental Methods,* Prentice Hall, Englewood Cliffs, NJ, 2003.
15. Bovey, F.A. and Mirau, P.A., *Nuclear Magnetic Resonance Spectroscopy,* 2nd ed., Academic Press, New York, 1988.
16. Harris, R.K. and Mann, B.E., *NMR and the Periodic Table,* Academic Press, London, 1978.
17. Nelson, J.H., *Nuclear Magnetic Resonance Spectroscopy,* 2nd ed., Wiley, New York, 2003.
18. Bruno, T.J. and Svoronos, P.D.N., *Handbook of Basic Tables for Chemical Analysis,* 2nd ed., CRC Press, Boca Raton, FL, 2003.

Alkanes

The chemical shift (in ppm) of C^i can be calculated from the following empirical equation

$$\Delta^i = -2.3 + \Sigma A_i$$

where ΣA_i is the sum of increments allowed for various substituents depending on their positions (α, β, γ, δ) relative to the ^{13}C in question, and -2.3 is the chemical shift for methane relative to tetramethylsilane (TMS).

^{13}C Chemical Shift Increments for A, the Shielding Term for Alkanes and Substituted Alkanes[9,10]

Substituent	Increments			
	α	β	γ	δ
>C– (sp³)	9.1	9.4	–2.5	0.3
>C=C< (sp²)	19.5	6.9	–2.1	0.4
C≡C– (sp)	4.4	5.6	–3.4	–0.6
C_6H_5	22.1	9.3	–2.6	0.3
–F	70.1	7.8	–6.8	0.0
–Cl	31.0	10.0	–5.1	–0.5
–Br	18.9	11.0	–3.8	–0.7
–I	–7.2	10.9	–1.5	–0.9
–OH	49.0	10.1	–6.2	0.0
–OR	49.0	10.1	–6.2	0.0
–CHO	29.9	–0.6	–2.7	0.0
–COR	22.5	3.0	–3.0	0.0
–COOH	20.1	2.0	–2.8	0.0
–COO⁻	24.5	3.5	–2.5	0.0
–COCl	33.1	2.3	–3.6	0.0
–COOR	22.6	2.0	–2.8	0.0
–OOCR	5.5	6.5	–6.0	—
–N<	28.3	11.3	–5.1	—
–NH_3^+	26.0	7.5	–4.6	0.0
[>N<]⁺	30.7	5.4	–7.2	–1.4
–ONO	54.3	6.1	–6.5	–0.5
–NO_2	61.6	3.1	–4.6	–1.0
–CON<	22.0	2.6	–3.2	–0.4
–NHCO–	31.3	8.3	–5.7	0.0
–C≡N	3.1	2.4	–3.3	–0.5
–NC	31.5	7.6	–3.0	0.0
–S–	10.6	11.4	–3.6	–0.4
–S–CO–	17.0	6.5	–3.1	0.0
–SO–	31.1	9.0	–3.5	0.0
–SO_2Cl	54.5	3.4	–3.0	0.0
–SCN	23.0	9.7	–3.0	0.0
–C(=S)N–	33.1	7.7	–2.5	0.6
–C=NOH(syn)	11.7	0.6	–1.8	0.0
–C=NOH(anti)	16.1	4.3	–1.5	0.0
$R_1 R_2 R_3Sn$ [a]	–5.2	4.0	–0.3	0.0

[a] R_1, R_2, and R_3 = organic substituents.

Thus, the C^{13} shift for C^i in 2-pentanol is predicted to be

$$\overset{\beta}{CH_3} - \overset{\alpha}{CH_2} - \overset{i}{CH_2} - \overset{\alpha'}{CH(OH)} - \overset{\beta'}{CH_3}$$

$$\delta^i = (-2.3) + \underset{\alpha \quad \beta \quad \alpha' \quad \beta' \quad OH}{[9.1 + 9.4 + 9.1 + 9.4 + 10.1]} = 44.8 \text{ ppm}$$

Alkenes

For a simple olefin of the type

$$\begin{matrix} \gamma & \beta & \alpha & i & & \alpha' & \beta' & \gamma' \\ -C & -C & -C & -C & = C & -C & -C & -C- \end{matrix}$$

$$\delta^i = 122.8 + \Sigma A_i$$

where: $A_\alpha = 10.6$, $A_\beta = 7.2$, $A_\gamma = -1.5$, $A_{\alpha'} = -7.9$, $A_{\beta'} = -1.8$, $A_{\gamma'} = 1.5$, and 122.8 is the chemical shift of the sp^2 carbon in ethene.

If the olefin is in the *cis-* configuration, an increment of -1.1 ppm must be added. Thus, the ^{13}C shift for C-3 in *cis*-3-hexene is predicted to be

$$\begin{matrix} \beta & \alpha & i & & \alpha' & \beta' \\ CH_3 & -CH_2 & -CH & =CH & -CH_2 & -CH_3 \end{matrix}$$

$$\delta^i = 122.8 + [\ 10.6 + 7.2 - 1.5 - 7.9\ ] + (-1.1) = 130.1 \text{ ppm}$$
$$\quad\quad\quad\quad (\alpha)\quad (\beta)\quad (\alpha')\quad (\beta')\quad\quad (cis)$$

Alkynes

For a simple alkyne of the type

$$\begin{matrix} \beta & \alpha & i & & \alpha' & \beta' \\ -C & -C & -C & \equiv C & -C & -C- \end{matrix}$$

$$\delta^i = 71.9 + \Sigma A_i$$

where increments A are given in the table below and 71.9 is the chemical shift of the sp carbon in acetylene.[9]

^{13}C Chemical-Shift Increments for A, the Shielding Term for Alkynes

Substituents	Increments			
	α	β	α'	β'
C (sp^3)	6.9	4.8	−5.7	2.3
−CH$_3$	7.0	—	−5.7	—
−CH$_2$CH$_3$	12.0	—	−3.5	—
−CH(CH$_3$)$_2$	16.0	—	−3.5	—
−CH$_2$OH	11.1	—	1.9	—
−COCH$_3$	31.4	—	4.0	—
−C$_6$H$_5$	12.7	—	6.4	—
−CH=CH$_2$	10.0	—	11.0	—
−Cl	−12.0	—	−15.0	—

Thus, the ^{13}C shift for C-A in 1-phenyl propyne is predicted to be

$$C_6H_5 - \underset{B}{C} \equiv \underset{A}{C} - CH_3$$

$$\delta^i = 71.9 + 7.0 + 6.4 = 85.3 \text{ ppm}$$

while the ^{13}C shift for C-B in the same compound is predicted to be

$$C_6H_5 - \underset{B}{C} \equiv \underset{A}{C^i} - CH_3$$

$$\delta^i = 71.9 + 12.7 - 5.7 = 78.9 \text{ ppm}$$

Benzenoid Aromatics

For a benzene derivative, C_6H_5–X, where X = substituent

$$\delta^i = 128.5 + \Sigma A_i$$

where ΣA_i is the sum of increments given below and 128.5 is the chemical shift of benzene.[9–10]

^{13}C Chemical Shift Increments for A, the Shielding Term for Benzenoid Aromatics

Substituent X[a]	Increments			
	C[i]	ortho	meta	para
–CH₃	9.3	0.8[9], 0.6[10]	0.0	–2.9[9], –3.1[10]
–CH₂CH₃	15.8[9], 15.7[10]	–0.4[9], –0.6[10]	–0.1	–2.6[9], –2.8[10]
–CH(CH₃)₂	20.3[9], 20.1[10]	–1.9[9], –2.0[10]	0.1[9], 0.0[10]	–2.4[9], –2.5[10]
–C(CH₃)₃	22.4[9], 22.1[10]	–3.1[9], –3.4[10]	–0.2[9], 0.4[10]	–2.9[9], –3.1[10]
–CH=CH₂	7.6	–1.8	–1.8	–3.5
–C≡CH	–6.1	3.8	0.4	–0.2
–C₆H₅	13.0	–1.1	0.5	–1.0
–CHO	8.6[9], 9.0[10]	1.3[9], 1.2[10]	0.6[9], 1.2[10]	5.5[9], 6.0[10]
–COCH₃	9.1[9], 9.3[10]	0.1[9], 0.2[10]	0.0[9], 0.2[10]	4.2
–CO₂H	2.1[9], 2.4[10]	1.5[9], 1.6[10]	0.0[9], –0.1[10]	5.1[9], 4.8[10]
–CO₂⁻	7.6	0.8	0.0	2.8
–CO₂R	2.1	1.2	0.0	4.4
–CONH₂	5.4	–0.3	–0.9	5.0
–CN	–15.4[9], –16.0[10]	3.6[9], 3.5[10]	0.6[9], 0.7[10]	3.9[9], 4.3[10]
–Cl	6.2[9], 6.4[10]	0.4[9], 0.2[10]	1.3[9], 10.0[10]	–1.9[9], –2.0[10]
–OH	26.9	–12.7	1.4	–7.3
–O–	39.6[10]	–8.2[10]	1.9[10]	–13.6[10]
–OCH₃	31.4[9], 30.2[10]	–14.4[9], –14.7[10]	1.0[9], 0.9[10]	–7.7[9], –8.1[10]
–OC₆H₅	29.1	–9.5	0.3	–5.3
–OC(=O)CH₃	23.0	–6.4	1.3	–2.3
–NH₂	18.7[9], 19.2[10]	–12.4	1.3	–9.5
–NHCH₃	21.7[10]	–16.2[10]	0.7[10]	–11.8[10]
–N(CH₃)₂	22.4	–15.7	0.8	–11.8
–NO₂	20.0[9], 19.6[10]	–4.8[9], –5.3[10]	0.9[9], 0.8[10]	5.8[9], 6.0[10]
–SH	2.2	0.7	0.4	–3.1
–SCH₃	9.9[10]	–2.0[10]	0.1[10]	–3.7[10]
–SO₃H	15.0	–2.2	1.3	3.8

[a] X–C₆H₅, where X = substituent.

As an example, the ^{13}C shift for the benzene carbon (C^i) carrying the carbonyl in 3,5-dinitro-acetophenone, $CH_3C(=O)(C_6H_3)(NO_2)_2$ is predicted to be

$$C^i = 128.5 + 9.1 + 2(0.9) = 132.4 \text{ ppm}$$

^{13}C NMR ABSORPTIONS OF MAJOR FUNCTIONAL GROUPS

The table below lists the ^{13}C chemical-shift ranges (in ppm) with the corresponding functional groups in descending order. A number of typical simple compounds for every family are given to illustrate the corresponding range. The shifts for the carbons of interest are given in parentheses, either for each carbon as it appears from left to right in the formula, or by the underscore.[1–15] Following the table, correlation charts depicting the ^{13}C chemical-shift ranges of various functional groups are presented. The expected peaks attributed to common solvents also appear in the correlation charts.

We provide a list of references that contain many of the ^{13}C chemical-shift ranges that appear below.[16–44] This list is certainly not complete and should be regularly updated. Reference 44 has a compilation of references for the various nuclei.

REFERENCES

1. Yoder, C.H. and Schaeffer, C.D., Jr., *Introduction to Multinuclear NMR: Theory and Application,* Benjamin/Cummings Publishing, Menlo Park, CA, 1987.
2. Brown, D.W., A short set of ^{13}C-NMR correlation tables, *J. Chem. Ed.,* 62, 209, 1985.
3. Silverstein, R.M. and Webster F.X., *Spectrometric Identification of Organic Compounds,* 6th ed., John Wiley and Sons, New York, 1998.
4. Becker, E.D., *High Resolution NMR, Theory and Chemical Applications,* 2nd ed., Academic Press, New York, 1980.
5. Gunther, H., *NMR Spectroscopy: Basic Principles, Concepts and Applications in Chemistry,* Wiley, New York, 2003.
6. Kitamaru, R., *Nuclear Magnetic Resonance: Principles and Theory,* Elsevier Science, New York, 1990.
7. Lambert, J.B., Holland, L.N., and Mazzola, E.P., *Nuclear Magnetic Resonance Spectroscopy: Introduction to Principles, Applications and Experimental Methods,* Prentice Hall, Englewood Cliffs, NJ, 2003.
8. Bovey, F.A. and Mirau, P.A., *Nuclear Magnetic Resonance Spectroscopy,* 2nd ed., Academic Press, New York, 1988.
9. Harris, R.K. and Mann, B.E., *NMR and the Periodic Table,* Academic Press, London, 1978.
10. Hore, P.J. and Hore, P.J., *Nuclear Magnetic Resonance,* Oxford University Press, Oxford, 1995.
11. Nelson, J.H., *Nuclear Magnetic Resonance Spectroscopy,* 2nd ed., Wiley, New York, 2003.
12. Levy, G. C., Lichter, R.L., and Nelson, G.L., *Carbon-13 Nuclear Magnetic Resonance Spectroscopy,* 2nd ed., Wiley, New York, 1980.
13. Pihlaja, K. and Kleinpeter, E., *Carbon-13 NMR Chemical Shifts in Structural and Stereochemical Analysis,* VCH, New York, 1994.
14. Aldrich Library of ^{1}H and ^{13}C FT-NMR Spectra, Aldrich Chemical Co., Milwaukee, 1996.
15. Bruno, T.J. and Svoronos, P.D.N., *Handbook of Basic Tables for Chemical Analysis,* 2nd ed., CRC Press, Boca Raton, FL, 2003.
16. *Adamantanes*
    Maciel, G.E., Dorn, H.C., Greene, R.L., Kleschick, W.A., Peterson, M.R., Jr., and Wahl, G.H., Jr., ^{13}C chemical shifts of monosubstituted adamantanes, *Org. Magn. Res.,* 6, 178, 1974.
17. *Amides*
    Jones, R.G. and Wilkins, J.M., Carbon-13 NMR spectra of a series of parasubstituted N,N-dimethylbenzamides, *Org. Magn. Res.,* 11, 20, 1978.
18. *Benzazoles*
    Sohr, P., Manyai, G., Hideg, K., Hankovszky, H., and Lex, L., Benzazoles, XIII: Determination of the E and Z configuration of isomeric 2-(2-benzimidazolyl)-di- and tetra-hydrothiophenes by IR, ^{1}H and ^{13}C NMR spectroscopy, *Org. Magn. Res.,* 14, 125, 1980.
19. *Carbazoles*
    Giraud, J. and Marzin, C., Comparative ^{13}C NMR study of deuterated and undeuterated dibenzothiophenes, dibenzofurans, carbazoles, fluorenes, and fluorenones, *Org. Magn. Res.,* 12, 647, 1979.

20. *Chlorinated Compounds*

Hawkes, G.E., Smith, R.A., and Roberts, J.D., Nuclear magnetic resonance spectroscopy: carbon-13 chemical shifts of chlorinated organic compounds, *J. Org. Chem.,* 39, 1276, 1974.

Mark, V. and Weil, E.D., The isomerization and chlorination of decachlorobi-2,4-cyclopentadien-1-yl, *J. Org. Chem.,* 36, 676, 1971.

21. *Diazoles and Diazines*

Faure, R., Vincent, E.J., Assef, G., Kister, J., and Metzger, J., Carbon-13 NMR study of substituent effects in the 1,3-diazole and -diazine series, *Org. Magn. Res.,* 9, 688, 1977.

22. *Disulfides*

Takata, T., Iida, K., and Oae, S., ^{13}C-NMR chemical shifts and coupling constants J_{C-H} of six-membered ring systems containing sulfur-sulfur linkage, *Het.,* 15, 847, 1981.

Bass, S.W. and Evans, S.A., Jr., Carbon-13 nuclear magnetic resonance spectral properties of alkyl disulfides, thiosulfinates, and thiosulfonates, *J. Org. Chem.,* 45, 710, 1980.

Freeman, F. and Angeletakis, C.N., Carbon-13 nuclear magnetic resonance study of the conformations of disulfides and their oxide derivatives, *J. Org. Chem.,* 47, 4194, 1982.

23. *Fluorenes and Fluorenones*

Giraud, J. and Marzin, C., Comparative ^{13}C NMR study of deuterated and undeuterated dibenzothiophenes, dibenzofurans, carbazoles, fluorenes and fluorenones, *Org. Magn. Res.,* 12, 647, 1979.

24. *Furans*

Giraud, H. and Marzin, C., Comparative ^{13}C NMR study of deuterated and undeuterated dibenzothiophenes, dibenzofurans, carbazoles, fluorenes and fluorenones, *Org. Magn. Res.,* 12, 647, 1979.

25. *Imines*

Allen, M. and Roberts, J.D., Effects of protonation and hydrogen bonding on carbon-13 chemical shifts of compounds containing the >C=N-group, *Can. J. Chem.,* 59, 451, 1981.

26. *Oxathianes*

Szarek, W.A., Vyas, D.M., Sepulchre, A.M., Gero, S.D., and Lukacs, G., Carbon-13 nuclear magnetic resonance spectra of 1,4-oxathiane derivatives, *Can. J. Chem.,* 52, 2041, 1974.

Murray, W.T., Kelly, J.W., and Evans, S.A., Jr., Synthesis of substituted 1,4-oxathianes: mechanistic details of diethoxytriphenylphosphorane- and triphenylphosphine/tetrachloromethane-promoted cyclodehydrations and ^{13}C NMR spectroscopy, *J. Org. Chem.,* 52, 525, 1987.

27. *Oximes*

Allen, M. and Roberts, J.D., Effects of protonation and hydrogen bonding on carbon-13 chemical shifts of compounds containing the >C=N-group, *Can. J. Chem.,* 59, 451, 1981.

28. *Polynuclear Aromatics (naphthalenes, anthracenes, pyrenes)*

Adcock, W., Aurangzeb, M., Kitching, W., Smith, N., and Doddzell, D., Substituent effects of carbon-13 nuclear magnetic resonance: concerning the π-inductive effect, *Aust. J. Chem.,* 27, 1817, 1974.

DuVernet, R. and Boekelheide, V., Nuclear magnetic resonance spectroscopy: ring-current effects on carbon-13 chemical shifts, *Proc. Nat. Acad. Sci., USA,* 71, 2961, 1974.

29. *Pyrazoles*

Puar, M.S., Rovnyak, G.C., Cohen, A.I., Toeplitz, B., and Gougoutas, J.Z., Orientation of the sulfoxide bond as a stereochemical probe: synthesis and ^{1}H and ^{13}C NMR of substituted thiopyrano[4,3-c]pyrazoles, *J. Org. Chem.,* 44, 2513, 1979.

30. *Sulfides*

Chauhan, M.S. and Still, I.W.J., ^{13}C nuclear magnetic resonance spectra of organic sulfur compounds: cyclic sulfides, sulfoxides, sulfones, and thiones, *Can. J. Chem.,* 53, 2880, 1975.

Gokel, G.W., Gerdes, H.M., and Dishong, D.M., Sulfur heterocycles, 3: heterogenous, phase-transfer, and acid-catalyzed potassium permanganate oxidation of sulfides to sulfones and a survey of their carbon-13 nuclear magnetic resonance spectra, *J. Org. Chem.,* 45, 3634, 1980.

Mohraz, M., Jiam-qi, W., Heilbronner, E., Solladie-Cavallo, A., and Matloubi-Moghadam, F., Some comments on the conformation of methyl phenyl sulfides, sulfoxides, and sulfones, *Helv. Chim. Acta,* 64, 97, 1981.

Srinivasan, C., Perumal, S., Arumugam, N., and Murugan, R., Linear free-energy relationship in naphthalene system-substituent effects on carbon-13 chemical shifts of substituted naphthyl-methyl sulfides, *Ind. J. Chem.,* 25A, 227, 1986.

31. Sulfites

Buchanan, G.W., Cousineau, C.M.E., and Mundell, T.C., Trimethylene sulfite conformations: effects of sterically demanding substituents at C-4,6 on ring geometry as assessed by ^{1}H and ^{13}C nuclear magnetic resonance, *Can. J. Chem.*, 56, 2019, 1978.

32. Sulfonamides

Chang, C., Floss, H.G., and Peck, G.E., Carbon-13 magnetic resonance spectroscopy of drugs: sulfonamides, *J. Med. Chem.*, 18, 505, 1975.

33. Sulfones (see also other families for the corresponding sulfones)

Fawcett, A.H., Ivin, K.J., and Stewart, C.D., Carbon-13 NMR spectra of monosulphones and disulphones: substitution rules and conformational effects, *Org. Magn. Res.*, 11, 360, 1978.

Gokel, G.W., Gerdes, H.M., and Dishong, D.M., Sulfur heterocycles, 3: heterogeneous, phase-transfer, and acid-catalyzed potassium permanganate oxidation of sulfides to sulfones and a survey of their carbon-13 nuclear magnetic resonance spectra, *J. Org. Chem.*, 45, 3634, 1980.

Balaji, T. and Reddy, D.B., Carbon-13 nuclear magnetic resonance spectra of some new arylcyclo-propyl sulphones, *Ind. J. Chem.*, 18B, 454, 1979.

34. Sulfoxides (see also other families for the corresponding sulfoxides)

Gatti, G., Levi, A., Lucchini, V., Modena, G., and Scorrano, G., Site of protonation in sulphoxides: carbon-13 nuclear magnetic resonance evidence, *J. Chem. Soc. Chem. Comm.*, 7, 251, 1973.

Harrison, C.R. and Hodge, P., Determination of the configuration of some penicillin S-oxides by ^{13}C nuclear magnetic resonance spectroscopy, *J. Chem. Soc., Perkin Trans. I*, 1772, 1976.

35. Sulfur Ylides

Matsuyama, H., Minato, H., and Kobayashi, M., Electrophilic sulfides (II) as a novel catalyst, V: Structure, nucleophilicity, and steric compression of stabilized sulfur ylides as observed by ^{13}C-NMR spectroscopy, *Bull. Chem. Soc. Japan*, 50, 3393, 1977.

36. Thianes

Willer, R.L. and Eliel, E.L., Conformational analysis, 34: Carbon-13 nuclear magnetic resonance spectra of saturated heterocycles, 6: methylthianes, *J. Am. Chem. Soc.*, 99, 1925, 1977.

Barbarella, G., Dembech, P., Garbesi, A., and Fara, A., ^{13}C NMR of organosulphur compounds, II: [13]C chemical shifts and conformational analysis of methyl-substituted thiacyclohexanes, *Org. Magn. Res.*, 8, 469, 1976.

Murray, W.T., Kelly, J.W., and Evans, S.A., Jr., Synthesis of substituted 1,4-oxathianes: mechanistic details of diethoxytriphenyl phosphorane and triphenylphosphine/tetrachloromethane-promoted cyclodehydrations and ^{13}C-NMR spectroscopy, *J. Org. Chem.*, 52, 525, 1987.

Block, E., Bazzi, A.A., Lambert, J.B., Wharry, S.M., Andersen, K.K., Dittmer, D.C., Patwardhan, B.H., and Smith, J.H., Carbon-13 and oxygen-17 nuclear magnetic resonance studies of orga-nosulfur compounds: the four-membered-ring-sulfone effect, *J. Org. Chem.*, 45, 4807, 1980.

Rooney, R.P. and Evans, S.A., Jr., Carbon-13 nuclear magnetic resonance spectra of trans-1-thiadecalin, trans-1,4-dithiadecalin, trans-1,4-oxathiadecalin, and the corresponding sulfoxides and sulfones, *J. Org. Chem.*, 45, 180, 1980.

37. Thiazines

Fronza, G., Mondelli, R., Scapini, G., Ronsisvalle, G., and Vittorio, F., ^{13}C NMR of N-heterocycles: conformation of phenothiazines and 2,3-diazaphenothiazines, *J. Magn. Res.*, 23, 437, 1976.

38. Thiazoles

Harrison, C.R. and Hodge, P., Determination of the configuration of some penicillin S-oxides by ^{13}C nuclear magnetic resonance spectroscopy, *J. Chem. Soc., Perkin Trans. I*, 16, 1772, 1976.

Chang, G., Floss, H.G., and Peck, G.E., Carbon-13 magnetic resonance spectroscopy of drugs: sulfonamides, *J. Med. Chem.*, 18, 505, 1975.

Elguero, J., Faure, R., Lazaro, R., and Vincent, E.J., ^{13}C NMR study of benzothiazole and its nitroderivatives, *Bull. Soc. Chim. Belg.*, 86, 95, 1977.

Faure, R., Galy, J.P., Vincent, E.J., and Elguero, J., Study of polyheteroaromatic pentagonal heterocycles by carbon-13 NMR: thiazoles and thiazolo[2,3-e]tetrazoles, *Can. J. Chem.*, 56, 46, 1978.

39. Thiochromanones

Chauhan, M.S. and Still, I.W.J., ^{13}C nuclear magnetic resonance spectra of organic sulfur com-pounds: cyclic sulfides, sulfoxides, sulfones and thiones, *Can. J. Chem.*, 53, 2880, 1975.

40. Thiones

Chauhan, M.S. and Still, I.W.J., ^{13}C nuclear magnetic resonance spectra of organic sulfur compounds: cyclic sulfides, sulfoxides, sulfones and thiones, *Can. J. Chem.,* 53, 2880, 1975.

41. Thiophenes

Perjessy, A., Janda, M., and Boykin, D.W., Transmission of substituent effects in thiophenes: infrared and carbon-13 nuclear magnetic resonance studies, *J. Org. Chem.,* 45, 1366, 1980.

Giraud, J. and Marzin, C., Comparative ^{13}C NMR study of deuterated and undeuterated dibenzothiophenes, dibenzofurans, carbazoles, fluorenes and flourenones, *Org. Magn. Res.,* 12, 647, 1979.

Clark, P.D., Ewing, D.F., and Scrowston, R.M., NMR studies of sulfur heterocycles, III: [13]C spectra of benzo[b]thiophene and the methylbenzo[b]thiophenes, *Org. Magn. Res.,* 8, 252, 1976.

Osamura, Y., Sayanagi, O., and Nishimoto, K., C-13 NMR chemical shifts and charge densities of substituted thiophenes: the effect of vacant do orbitals, *Bull. Chem. Soc. Japan,* 49, 845, 1976.

Balkau, F., Fuller, M.W., and Heffernan, M.L., Deceptive simplicity in ABMX NMR spectra, I: dibenzothiophen and 9.9′-dicarbazyl, *Aust. J. Chem.,* 24, 2293, 1971.

Geneste, P., Olive, J.L., Ung, S.N., El Faghi, M.E.A., Easton, J.W., Beierbeck, H., and Saunders, J.K., Carbon-13 nuclear magnetic resonance study of benzo[b]thiophenes and benzo[b]-thiophene S-oxides and S,S-dioxides, *J. Org. Chem.,* 44, 2887, 1979.

Benassi, R., Folli, U., Iarossi, D., Schenetti, L., and Tadei, F., Conformational analysis of organic carbonyl compounds, Part 3. A ^{1}H and ^{13}C nuclear magnetic resonance study of formyl and acetyl derivatives of benzo[b]thiophen, *J. Chem. Soc., Perkin Trans. II,* 7, 911, 1983.

Kiezel, L., Liszka, M., and Rutkowski, M., Carbon-13 magnetic resonance spectra of benzothiophene and dibenzothiophene, *Spec. Lett.,* 12, 45, 1979.

Fujieda, K., Takahashi, K., and Sone, T., The C-13 NMR spectra of thiophenes, II: 2-substituted thiophenes, *Bull. Chem. Soc. Japan,* 58, 1587, 1985.

Satonaka, H. and Watanabe, M., NMR spectra of 2-(2-nitrovinyl) thiophenes, *Bull. Chem. Soc. Jap.,* 58, 3651, 1985.

Stuart, J.G., Quast, M.J., Martin, G.E., Lynch, V.M., Simmonsen, H., Lee, M.L., Castle, R.N., Dallas, J.L., John B.K., and Johnson, L.R.F., Benzannelated analogs of phenanthro [1,2-b]-[2,1-b]thiophene: synthesis and structural characterization by two-dimensional NMR and X-ray techniques, *J. Heterocyclic Chem.,* 23, 1215, 1986.

42. Thiopyrans

Senda, Y., Kasahara, A., Izumi, T., and Takeda, T., Carbon-13 NMR spectra of 4-chromanone, 4H-1-benzothiopyran-4-one, 4H-1-benzothiopyran-4-one 1,1-dioxide, and their substituted homologs, *Bull. Chem. Soc. Japan,* 50, 2789, 1977.

43. Thiosulfinates and Thiosulfonates

Bass, S.W. and Evans, S.A., Jr., Carbon-13 nuclear magnetic resonance spectral properties of alkyl disulfides, thiosulfinates, and thiosulfonates, *J. Org. Chem.,* 45, 710, 1980.

44. University of Wisconsin, NMR Bibliography; available on-line at http://www.chem.wisc.edu/areas/reich/Handouts/nmr/NMR-Biblio.htm.

^{13}C NMR Chemical-Shift Ranges of Organic Functional Groups

δ, ppm	Group	Family	Example (δ of underlined carbon)	
220–165	>C=O	Ketones	$(CH_3)_2\underline{C}O$	(206.0)
			$(CH_3)_2CH\underline{C}OCH_3$	(212.1)
		Aldehydes	$CH_3\underline{C}HO$	(199.7)
		α,-Unsaturated carbonyls	$CH_3CH=CH\underline{C}HO$	(192.4)
			$CH_2=CH\underline{C}OCH_3$	(169.9)
		Carboxylic acids	$H\underline{C}O_2H$	(166.0)
			$CH_3\underline{C}O_2H$	(178.1)
		Amides	$H\underline{C}ONH_2$	(165.0)
			$CH_3\underline{C}ONH_2$	(172.7)
		Esters	$CH_3\underline{C}O_2CH_2CH_3$	(170.3)
			$CH_2=CH\underline{C}O_2CH_3$	(165.5)
140–120	>C=C<	Aromatic alkenes	C_6H_6	(128.5)
			$CH_2=CH_2$	(123.2)
			$CH_2=\underline{C}HCH_3$	(115.9, 136.2)
			$CH_2=\underline{C}HCH_2Cl$	(117.5, 133.7)
			$CH_3CH=\underline{C}HCH_2CH_3$	(132.7)
125–115	–C≡N	Nitriles	$CH_3–\underline{C}≡N$	(117.7)
80–70	–C≡C–	Alkynes	$H\underline{C}≡CH$	(71.9)
			$CH_3\underline{C}≡CCH_3$	(73.9)
70–45	>Ç–O	Esters, alcohols	$\underline{C}H_3OOCCH_2CH_3$	(57.6, 67.9)
			$HO\underline{C}H_3$	(49.0)
			$HO\underline{C}H_2CH_3$	(57.0)
40–20	>Ç–NH$_2$	Amines	$\underline{C}H_3NH_2$	(26.9)
			$CH_3\underline{C}H_2NH_2$	(35.9)
30–15	–S–CH$_3$	Sulfides (thioethers)	$C_6H_5–S–\underline{C}H_3$	15.6
30–(–2.3)	>CH–	Alkanes, cycloalkanes	$\underline{C}H_4$	(–2.3)
			$\underline{C}H_3CH_3$	(5.7)
			$\underline{C}H_3\underline{C}H_2CH_3$	(15.8, 16.3)
			$\underline{C}H_3\underline{C}H_2CH_2CH_3$	(13.4, 25.2)
			$\underline{C}H_3\underline{C}H_2\underline{C}H_2CH_2CH_3$	(13.9, 22.8, 34.7)
			cyclohexane	(26.9)

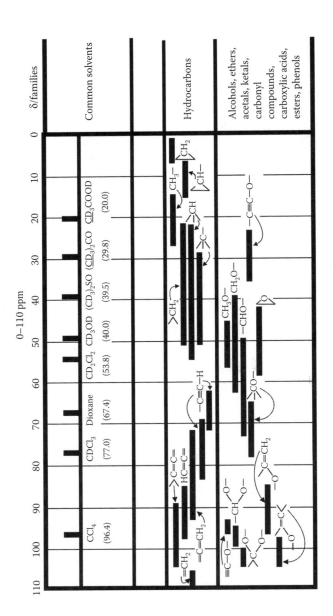

Figure 3.29

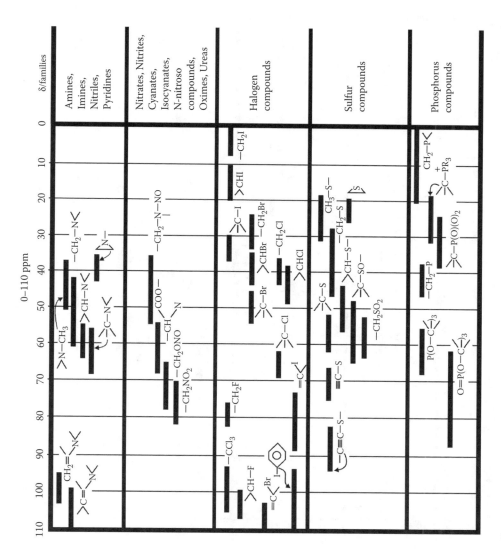

Figure 3.30

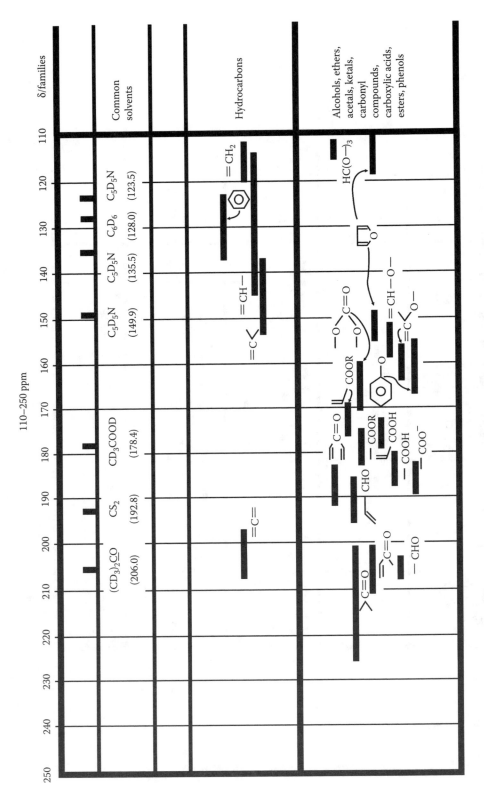

Figure 3.31

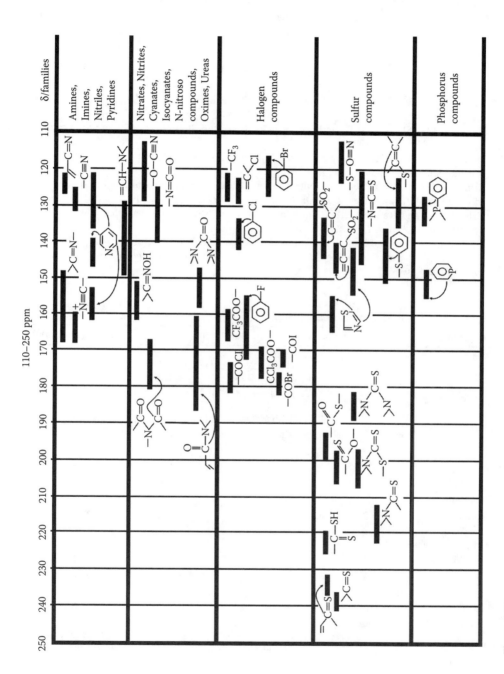

Figure 3.32

^{15}N CHEMICAL SHIFTS FOR COMMON STANDARDS

The following table lists the ^{15}N chemical shifts (in ppm) for common standards. The estimated uncertainty is less than 0.1 ppm. Nitromethane (according to Levy and Lichter[1]) is the most suitable primary measurement reference, but it has the disadvantage of lying in the low-field end of the spectrum. Thus, ammonia (which lies in the most up-field region) is the most suitable for routine experimental use.[1-7]

REFERENCES

1. Levy, G.C. and Lichter, R.L., *Nitrogen-15 Nuclear Magnetic Resonance Spectroscopy,* John Wiley and Sons, New York, 1979.
2. Lambert, J.B., Shurvell, H.F., Verbit, L., Cooks, R.G., and Stout, G.H., *Organic Structural Analysis,* Macmillan, New York, 1976.
3. Witanowski, M., Stefaniak, L., Szymanski, S., and Januszewski, H., External neat nitromethane scale for nitrogen chemical shifts, *J. Magn. Res.,* 28, 217, 1977.
4. Srinivasan, P.R. and Lichter, R.L., Nitrogen-15 nuclear magnetic resonance spectroscopy: evaluation of chemical shift references, *J. Magn. Res.,* 28, 227, 1977.
5. Briggs, J.M. and Randall, E.W., Nitrogen-15 chemical shifts in concentrated aqueous solutions of ammonium salts, *Mol. Phys.,* 26, 699, 1973.
6. Becker, E.D., Proposed scale for nitrogen chemical shifts, *J. Magn. Res.,* 4, 142, 1971.
7. Bruno, T.J. and Svoronos, P.D.N., *Handbook of Basic Tables for Chemical Analysis,* 2nd ed., CRC Press, Boca Raton, FL, 2003.

^{15}N Chemical Shifts for Common Standards

Compound	Formula	Conditions	Chemical Shift, ppm
Ammonia	NH_3	Vapor (0.5 MPa)	−15.9
		Liquid (25°C), anhydrous	0.0
		Liquid (−50°C)	3.37
Ammonium chloride	NH_4Cl	2.9M (in 1M HCl)	24.93
		1.0M (in 10M HCl)	30.31
		Aqueous solution (saturated)	27.34
Tetraethylammonium chloride	$(C_2H_5)_4N^+Cl^-$	Aqueous solution (saturated)	43.54
		Chloroform solution (saturated)	45.68
		Aqueous solution (0.3M)	63.94
		Aqueous solution (saturated)	64.39
		Chloroform solution (0.075M)	65.69
Tetramethyl urea	$[(CH_3)_2N]_2CO$	Neat	62.50
Dimethylformamide (DMF)	$(CH_3)_2NCHO$	Neat	103.81
Nitric acid (aqueous solution)	HNO_3	1M	375.80
		2M	367.84
		9M	365.86
		10M	362.00
		15.7M	348.92
Sodium nitrate	$NaNO_3$	Aqueous solution (saturated)	376.53
Ammonium nitrate	NH_4NO_3	Aqueous solution (saturated)	376.25
		5M (in 2M HNO_3)	375.59
		4M (in 2M HNO_3)	374.68
Nitromethane	CH_3NO_2	1:1 (vol/vol) in $CDCl_3$	379.60
		0.03M Cr(acac)$_3$	—
		Neat	380.23

15N CHEMICAL SHIFTS OF MAJOR FUNCTIONAL GROUPS

The following correlation chart contains ^{15}N chemical shifts of various organic nitrogen compounds. Chemical shifts are often expressed relative to different standards (NH_3, NH_4Cl, CH_3NO_2, NH_4NO_3, HNO_3, etc.) and are interconvertible.

In view of the large chemical-shift range (up to 900 ppm), caution in using these correlation charts is of great importance, as the chemical shifts are greatly dependent on the inductive, mesomeric, or hybridization effects of the neighboring groups, as well as the solvent used.

Chemical shifts are sensitive to hydrogen bonding and are solvent dependent, as seen in case of pyridine. Consequently, the reference as well as the solvent should always accompany chemical-shift data. No data are given on peptides and other biochemical compounds. All shifts given in these correlation charts are relative to ammonia unless otherwise specified. A section of "miscellaneous" data gives the chemical shift of special compounds relative to unusual standards.[1-16] Reference 17 contains a compilation of publications that involve various nuclei.

REFERENCES

1. Levy, G.C. and Lichter, R.L., *Nitrogen-15 Nuclear Magnetic Resonance Spectroscopy*, John Wiley and Sons, New York, 1979.

2. Yoder, C.H. and Schaeffer, C.D., Jr., *Introduction to Multinuclear NMR*, Benjamin/Cummings, Menlo Park, CA, 1987.

3. Duthaler, R.O. and Roberts, J.D., Effects of solvent, protonation, and N-alkylation on the ^{15}N chemical shifts of pyridine and related compounds, *J. Am. Chem. Soc.*, 100, 4969, 1978.

4. Duthaler, R.O. and Roberts, J.D., Steric and electronic effects on ^{15}N chemical shifts of saturated aliphatic amines and their hydrochlorides, *J. Am. Chem. Soc.*, 100, 3889, 1978.

5. Kozerski, L. and von Philipsborn, W., ^{15}N chemical shifts as a conformational probe in enaminones: a variable temperature study at natural isotope abundance, *Org. Magn. Res.*, 17, 306, 1981.

6. Duthaler, R.O. and Roberts, J.D., Steric and electronic effects on ^{15}N chemical shifts of piperidine and decahydroquinoline hydrochlorides, *J. Am. Chem. Soc.*, 100, 3882, 1978.

7. Duthaler, R.O. and Roberts, J.D., Nitrogen-15 nuclear magnetic resonance spectroscopy: solvent effects on the ^{15}N chemical shifts of saturated amines and their hydrochlorides, *J. Magn. Res.*, 34, 129, 1979.

8. Psota, L., Franzen-Sieveking, M., Turnier, J., and Lichter, R.L., Nitrogen nuclear magnetic resonance spectroscopy: nitrogen-15 and proton chemical shifts of methylanilines and methylanilinium ions, *Org. Magn. Res.*, 11, 401, 1978.

9. Subramanian, P.K., Chandra Sekara, N., and Ramalingam, K., Steric effects on nitrogen-15 chemical shifts of 4-aminooxanes (tetrahydropyrans), 4-amino-thianes, and the corresponding N,N-dimethyl derivatives: use of nitrogen-15 shifts as an aid in stereochemical analysis of these heterocyclic systems, *J. Org. Chem.*, 47, 1933, 1982.

10. Schuster, I.I. and Roberts, J.D., Proximity effects on nitrogen-15 chemical shifts of 8-substituted 1-nitronaphthalenes and 1-naphthylamines, *J. Org. Chem.*, 45, 284, 1980.

11. Kupce, E., Liepins, E., Pudova, O., and Lukevics, E., Indirect nuclear spin-spin coupling constants of nitrogen-15 to silicon-29 in silylamines, *J. Chem. Soc., Chem. Comm.*, 9, 581, 1984.

12. Allen, M. and Roberts, J.D., Effects of protonation and hydrogen bonding on nitrogen-15 chemical shifts of compounds containing the >C=N–group, *J. Org. Chem.*, 45, 130, 1980.

13. Brownlee, R.T.C. and Sadek, M., Natural abundance ^{15}N chemical shifts in substituted benzamides and thiobenzamides, *Magn. Res. Chem.*, 24, 821, 1986.

14. Dega-Szafran, Z., Szafran, M., Stefaniak, L., Brevard, C., and Bourdonneau, M., Nitrogen-15 nuclear magnetic resonance studies of hydrogen bonding and proton transfer in some pyridine trifluoroacetates in dichloromethane, *Magn. Res. Chem.*, 24, 424, 1986.

15. Lambert, J.B., Shurvell, H.F., Verbit, L., Cooks, R.G., and Stout, G.H., *Organic Structural Analysis*, Macmillan, New York, 1976.

16. Bruno, T.J. and Svoronos, P.D.N., *Handbook of Basic Tables for Chemical Analysis*, 2nd ed., CRC Press, Boca Raton, FL, 2003.

17. University of Wisconsin, NMR Bibliography; available on-line at http://ww.chem.wisc.edu/areas/reich/Handouts/nmr/NMR-Biblio.htm.

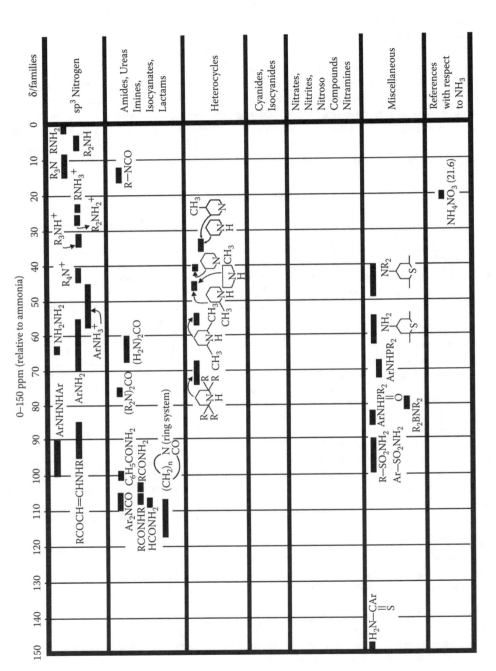

Figure 3.33

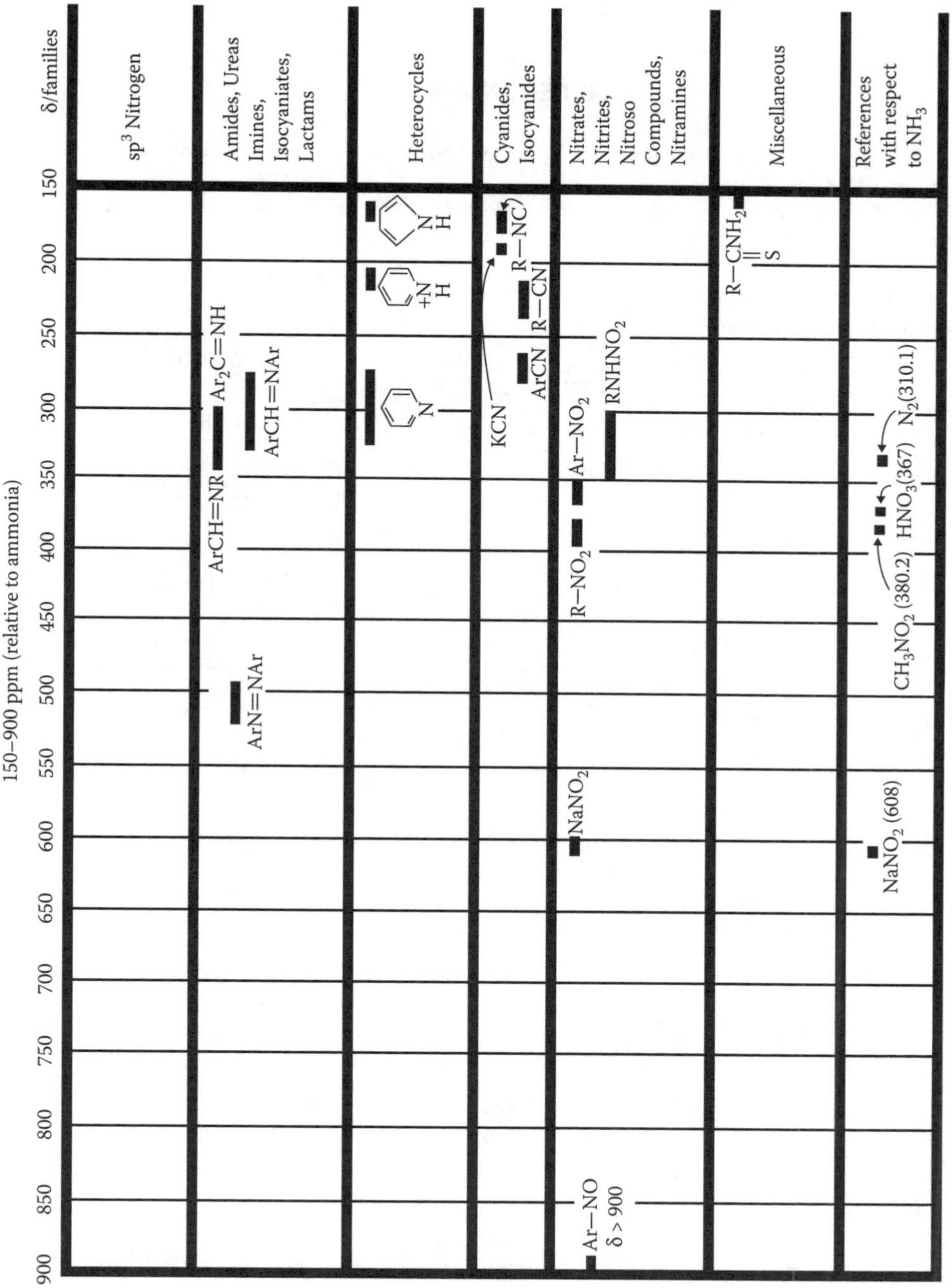
150–900 ppm (relative to ammonia)
δ/families
sp^3 Nitrogen
Amides, Ureas Imines, Isocyaniates, Lactams
Heterocycles
Cyanides, Isocyanides
Nitrates, Nitrites, Nitroso Compounds, Nitramines
Miscellaneous
References with respect to NH_3
ArCH=NR
Ar_2C=NH
ArCH=NAr
ArN=NAr
KCN
R—NC
R—CN
ArCN
Ar—NO_2
R—NO_2
$RNHNO_2$
$NaNO_2$
Ar—NO
δ > 900
R—CNH_2
S
CH_3NO_2 (380.2)
HNO_3(367)
N_2(310.1)
$NaNO_2$ (608)

Figure 3.34

SPIN-SPIN COUPLING TO ^{15}N

The following four tables provide representative spin-spin coupling values (J_{NH}, Hz) to ^{15}N. When consulting this table, the reader should keep in mind the considerable mesomeric, inductive, and hybridization effects of the neighboring groups.[1-2]

REFERENCES

1. Levy, G.C. and Lichter, R.L., *Nitrogen-15 Nuclear Magnetic Resonance Spectroscopy*, John Wiley and Sons, New York, 1979.
2. Bruno, T.J. and Svoronos, P.D.N., *Handbook of Basic Tables for Chemical Analysis*, 2nd ed., CRC Press, Boca Raton, FL, 2003.

$^{15}N–^{1}H$ Coupling Constants

Bond Type	Family	J_{NH}, Hz	Example
One bond	Ammonia	−61.2	NH_3
	Amines, aliphatic (1°, 2°)	≈−65	CH_3NH_2 (−64.5)
			$(CH_3)_2NH$ (−67.0)
	Ammonium salts	≈−75	CH_3NH_3Cl (−75.4)
			$(CH_3)_2NH_2Cl$ (−76.1)
			$C_6H_5NH_3^+$ (−76)
	Amines, aromatic (1°, 2°)	−78 to −95	$C_6H_5NH_2$ (−78.5)
			p-$CH_3O–C_6H_4–NH_2$ (−79.4)
			p-$O_2N–C_6H_4–NH_2$ (−92.6)
	Sulfonamides	≈−80	$C_6H_5SO_2NH_2$ (−80.8)
	Hydrazines	−90 to −100	$C_6H_5NHNH_2$ (−89.6)
	Amides (1°, 2°)	−85 to −95	$HCONH_2$ (−88) (syn); (−92) (anti)
	Pyrroles	−95 to −100	Pyrrole (−96.53)
	Nitriles, salts	≈−135	$CH_3C{\equiv}NH^+$ (−136)
Two bonds	Amines	≈−1	CH_3NH_2 (−1.0)
			$(CH_3)_3N$ (−0.85)
	Pyridinium salts	≈−3	$C_5H_5NH^+$ (−3)
	Pyrroles	≈−5	C_4H_4NH (−4.52)
	Thiazoles	≈−10	C_3H_3NS (−10)
	Pyridines	≈−10	C_5H_5N (−10.76)
	Oximes, syn	≈−15	—
	Oximes, anti	−2.5 to 2.5	—
Three bonds	Nitriles, salts	≈2–4	$CH_3C{\equiv}NH^+$ (2.8)
	Amides	≈1–2	CH_3CONH_2 (1.3)
	Anilines	≈1–2	$C_6H_5NH_2$ (1.5, 1.8)
	Pyridines	≈3.1	C_5H_5N (0.2)
	Nitriles	−1 to −2	$CH_3C{\equiv}N$ (−1.7)
	Pyridinium salts	≈−4	$C_5H_5NH^+$ (−3.98)
	Pyrroles	≈−5	C_4H_4NH (−5.39)

$^{15}N-^{13}C$ Coupling Constants

Bond Type	Family	J_{CN}, Hz	Example
One bond	Amines, aliphatic	≈-4	CH_3NH_2 (−4.5)
			$CH_3(CH_2)_2NH_2$ (−3.9)
	Ammonium salts (aliphatic)	≈-5	$CH_3(CH_2)_2NH_3^+$ (−4.4)
	Ammonium salts (aromatic)	≈-9	$C_6H_5NH_3^+$ (−8.9)
	Pyrroles	≈-10	C_4H_4NH (−10.3)
	Amines, aromatic	−11 to −15	$C_6H_5NH_2$ (−11.43)
	Nitro compounds	−10 to −15	CH_3NO_2 (−10.5)
			$C_6H_5NO_2$ (−14.5)
	Nitriles	≈-17	$CH_3C{\equiv}N$ (−17.5)
	Amides	≈-14	$C_6H_5NHCOCH_3$ (−14.3) (CO);
			(−14.1) (C_1)
Two bonds	Amides	7–9	CH_3CONH_2 (9.5)
	Nitriles	≈3	$CH_3C{\equiv}N$ (3.0)
	Pyridines and N-pyridinium salts	$\approx1-3$	C_5H_5N (2.53)
			$C_5H_5NH^+$ (2.01)
			C_5H_5NO (1.43)
	Amines, aliphatic	$\approx1-2$	$CH_3CH_2CH_2NH_2$ (1.2)
	Nitro compounds, aromatic	≈-1 to −2	$C_6H_5NO_2$ (−1.67)
	Amines, aromatic	≈-1 to −2	$C_6H_5NH_2$ (−2.68)
			$C_6H_5NH_3^+$ (−1.5)
	Pyrroles	≈-4	C_4H_4NH (−3.92)
Three bonds	Amides	9	$CH_2{=}CHCONH_2$ (19)
	Ammonium salts	1–9	$CH_3(CH_2)_2NH_3^+$ (1.3)
			$C_6H_5NH_3^+$ (2.1)
	Pyridines	≈3	C_5H_5N (2.53)
	Amines, aromatic	≈-1 to −3	$C_6H_5NH_2$ (−2.68)
	Nitro compounds	≈-2	$C_6H_5NO_2$ (−1.67)
	Pyrroles	≈-4	C_4H_4NH (−3.92)

$^{15}N-^{15}N$ Coupling Constants

Family	J_{NN}, Hz	Example
Azocompounds	12–25	$C_6H_5N{=}NC(CH_3)_2C_6H_5$ anti (17); syn (21)
N-nitrosamines	≈19	$(C_6H_5CH_2)_2N{-}N{=}O$ (19)
Hydrazones	≈10	$p{-}O_2NC_6H_4CH{=}N{-}NHC_6H_5$ (10.7)
Hydrazines	≈7	$C_6H_5NHNH_2$ (6.7)

$^{15}N-^{19}F$ Coupling Constants

Family	J_{NF}, Hz	Example
Difluorodiazines		
trans	≈190 ($^1J_{NF}$)	F−N=N−F (190)
	≈102 ($^2J_{NF}$)	F−N=N−F (102)
cis	≈203 ($^1J_{NF}$)	F−N=N−F (203)
	≈52 ($^2J_{NF}$)	F−N=N−F (52)
Fluoropyridines		
2-fluoro-	—	−52.5
3-fluoro-	—	+3.6
Fluoroanilines		
2-fluoro-	0	$1,2{-}C_6H_4F(NH_2)$
3-fluoro-	0	$1,3{-}C_6H_4F(NH_2)$
4-fluoro-	1.5	$1,4{-}C_6H_4F(NH_2)$
Fluoroanilinium salts		
2-fluoro-	1.4	$1,2{-}C_6H_4F(NH_3^+)$
3-fluoro-	0.2	$1,3{-}C_6H_4F(NH_3^+)$
4-fluoro-	0	$1,4{-}C_6H_4F(NH_3^+)$

^{19}F CHEMICAL-SHIFT RANGES

The following table lists the ^{19}F chemical-shift ranges (in ppm) relative to neat $CFCl_3$.[1-4]

REFERENCES

1. Yoder, C.H. and Schaeffer, C.D., Jr., *Introduction to Multinuclear NMR: Theory and Application,* Benjamin/Cummings, Menlo Park, CA, 1987.
2. Dungan, C.H. and Van Wazer, I.R., *Compilation of Reported ^{19}F Chemical Shifts 1951 to Mid 1967,* Wiley Interscience, New York, 1970.
3. Emsley, J.W., Phillips, L., and Wray, V., *Fluorine Coupling Constants,* Pergamon, New York, 1977.
4. Bruno, T.J. and Svoronos, P.D.N., *Handbook of Basic Tables for Chemical Analysis,* 2nd ed., CRC Press, Boca Raton, FL, 2003.

^{19}F Chemical-Shift Ranges Relative to Neat $CFCl_3$, ppm

Compound Type	Chemical Shift Range Relative to Neat $CFCl_3$, ppm
F–C(=O)	−70 to −200
–CF$_3$	+40 to +80
–CF$_2$–	+80 to +140
>CF–	+140 to +250
Ar–F[a]	+80 to +170

[a]Ar = aromatic moiety.

¹⁹F CHEMICAL-SHIFT RANGES OF SOME FLUORINE-CONTAINING COMPOUNDS

The following correlation charts list the ¹⁹F chemical shifts of some fluorine-containing compounds relative to neat $CFCl_3$. All chemical shifts are those of neat samples, and the values pertain to the fluorine present in the molecule.[1-4]

REFERENCES

1. Dungan, C.H. and Van Wazer, I.R., *Compilation of Reported ¹⁹F Chemical Shifts 1951 to Mid 1967*, Wiley Interscience, New York, 1970.
2. Emsley, J.W., Phillips, L., and Wray, V., *Fluorine Coupling Constants*, Pergamon, New York, 1977.
3. Bruno, T.J. and Svoronos, P.D.N., *Handbook of Basic Tables for Chemical Analysis*, 2nd ed., CRC Press, Boca Raton, FL, 2003.
4. University of Wisconsin, NMR Bibliography; available on-line at http://www.chem.wisc.edu/areas/reich/Handouts/nmr/NMR-Biblio.htm.

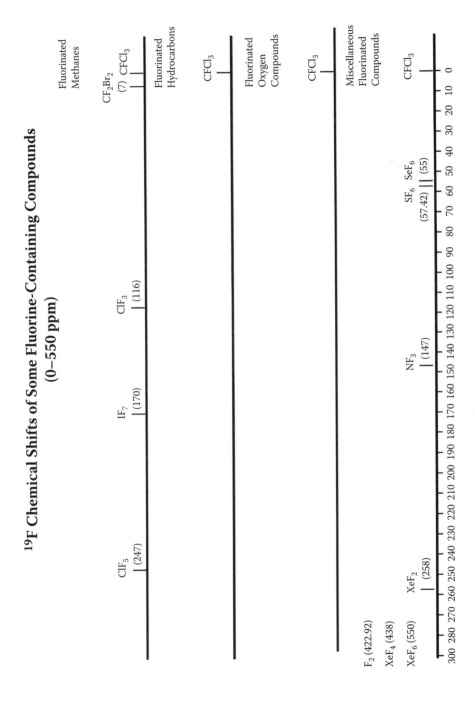

Figure 3.35

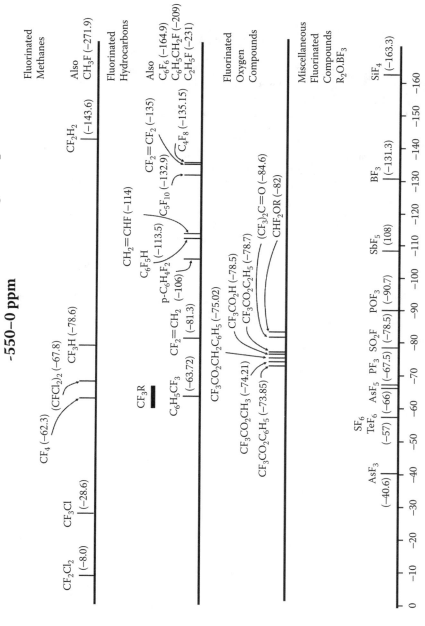

Figure 3.36

FLUORINE COUPLING CONSTANTS

The following tables give the most important fluorine coupling constants, namely, J_{FN}, J_{FCF}, and J_{CF}, together with some typical examples.[1-10] The coupling-constant values vary with the solvent used.[3] The book by Emsley, Phillips, and Wray[1] gives a complete, detailed list of various specific compounds.

REFERENCES

1. Emsley, J.W., Phillips, L., and Wray, V., *Fluorine Coupling Constants,* Pergamon Press, Oxford, 1977.
2. Lambert, J.R., Shurvell, H.F., Verbit, L., Cooks, R.G., and Stout, G.H., *Organic Structural Analysis,* Macmillan, New York, 1976.
3. Yoder, C.H. and Schaeffer, C.D., Jr., *Introduction to Multinuclear NMR: Theory and Application,* Benjamin/Cummings, Menlo Park, CA, 1987.
4. Schaeffer, T., Marat, K., Peeling, J., and Veregin, R.P., Signs and mechanisms of ^{13}C: ^{19}F spin-spin coupling constants in benzotrifluoride and its derivatives, *Can. J. Chem.,* 61, 2779, 1983.
5. Adcock, W. and Kok, G.B., Polar substituent effects on ^{19}F chemical shifts of aryl and vinyl fluorides: a fluorine-19 nuclear magnetic resonance study of some 1,1-difluoro-2-(4-substituted-bicyclo[2,2,2]oct-1-yl)ethenes, *J. Org. Chem.,* 50, 1079, 1985.
6. Newmark, R.A. and Hill, J.R., Carbon-13-fluorine-19 coupling constants in benzotrifluorides, *Org. Magn. Res.,* 9, 589, 1977.
7. Adcock, W. and Abeywickrema, A.N., Concerning the origin of substituent-induced fluorine-19 chemical shifts in aliphatic fluorides: carbon-13 and fluorine-19 nuclear magnetic resonance study of 1-fluoro-4-phenylbicyclo[2,2,2]octanes substituted in the arene ring, *J. Org. Chem.,* 47, 2945, 1982.
8. Dungan, C.H. and Van Wazer, I.R., *Compilation of Reported ^{19}F Chemical Shifts 1951 to Mid 1967,* Wiley Interscience, New York, 1970.
9. Emsley, J.W., Phillips, L., and Wray, V., *Fluorine Coupling Constants,* Pergamon, New York, 1977.
10. Bruno, T.J. and Svoronos, P.D.N., *Handbook of Basic Tables for Chemical Analysis,* 2nd ed., CRC Press, Boca Raton, FL, 2003.

$^{19}F–^1H$ Coupling Constants

Fluorinated Family	J_{FH}, Hz	Example
		Two Bonds
Alkanes	45–80	CH_3F (45); CH_2F_2 (50); CHF_3 (79); C_2H_5F (47); CH_3CHF_2 (57); CH_2FCH_2F (48); CH_2FCHF_2 (54); CF_3CH_2F (45); $CF_2HCF_2CF_3$ (52)
Alkyl chlorides	49–65	FCl_2CH (53); CF_2HCl (63); $FCHCl–CHCl_2$ (49); $FCH_2–CH_2Cl$ (46)
Alkyl bromides	45–50	$FBrCHCH_3$ (50.5); $FH_2C–CH_2Br$ (46); $FBrCH-CHFBr$ (49)
Alkenes	45–80	$FHC=CHF$ (cis-71.7; trans-75.1); $CH_2=CHF$ (85); $CF_2=CHF$ (70.5); $FCH_2CH=CH_2$ (47.5)
Aromatics	45–75	$Cl–C_6H_4–CH_2F$ (m-47, p-48); $FH_2C–C_6H_4–NO_2$ (m-47, p-48); $FH_2C–C_6H_4–F$ (m-48, p-48); $p-Br–C_6H_4–OCF_2H$ (73)
Ethers	40–75	FH_2COCH_3 (74); $CF_2HCF_2OCH_3$ (46); $F_2HC–O–CH(CH_3)_2$ (75)
Ketones	45–50	FCH_2COCH_3 (47); $F_2HC–COCH_3$ (54); $CH_3CH_2CHFCOCH_3$ (50); $F_2HC–COCH(CF_3)_2$ (54)
Aldehydes	≈50	$CH_3CH_2CHFCHO$ (51)
Esters	45–70	$CFH_2CO_2CH_2CH_3$ (47); $CH_3CHFCO_2CH_2CH_3$ (48)
		Three Bonds
Alkanes	2–25	CF_2HCH_3 (21); $(CH_3)_3CF$ (20.4); $CH_3CHFCH_2CH_2CH_3$ (23); CF_3CH_3 (13)
Alkyl chlorides	8–20	CF_2HCHCl_2 (8); CF_2ClCH_3 (15)
Alkyl bromides	15–25	CF_2BrCH_2Br (22); CF_2BrCH_3 (16); $FC(CH_3)_2CHBrCH_3$ (21)
Alkenes	(−5)–60 J_{HCF} (cisoid) < 20 J_{HCF} (transoid) > 20	$CHF=CHF$ (cis-19.6; trans-2.8); $CH_2=CHF$ (cis-19.6; trans-51.8); $CHF=CF_2$ (cis-(−4.2); trans-12.5); $CH_2=CF_2$ (cis-0.6; trans-33.8)
Alcohols	5–30	CF_3CH_2OH (8); FCH_2CH_2OH (29); CH_3CHFCH_2OH (23.6, 23.6); $CF_3CH(OH)CH_3$ (7.5); $CF_3CH(OH)CF_3$ (6); $FC(CH_3)_2C(OH)(CH_3)_2$ (23)
Ketones	5–25	$CH_3CH_2CHFCOCH_3$ (24); $FC(CH_3)_2COCH_3$ (21); $(CF_3)_2CHCOCH_3$ (8); $CF_2HCOCH(CF_3)_2$ (7)
Aldehydes	10–25	$(CH_3)_2CFCHO$ (22)
Esters	10–25	$CH_3CHFCO_2CH_2CH_3$ (23); $(CH_3CH_2)_2CFCO_2CH_3$ (16.5)

$^{19}F-^{19}F$ Coupling Constants

Carbon	J_{FCF}, Hz	Examples
Two Bonds		
Saturated (sp³)	140–250	$CF_3CF_2{}^{a,b}CFHCH_3$ ($J_{ab} = 270$); $CF_2{}^{a,b}BrCHFSO_2F$ ($J_{ab} = 188$); $CH_3O-CF_2{}^{a,b}CFHSO_2F$ ($J_{ab} = 147$); $CH_3O-CF_2{}^{a,b}$; $CFHCl$ ($J_{ab} = 142$); $CH_3S-CF_2{}^{a,b}CFHCl$ ($J_{ab} = 222$)
Cycloalkanes	150–240	$F_2C(CH_2)_2$ (150) (3-membered); $F_2C(CH_2)_3$ (200) (4-membered); $F_2C(CH_2)_4$ (240) (5-membered); $F_2C(CH_2)_5$ (228) (6-membered)
Unsaturated (sp²)	≤100	$CF_2=CH_2$ (31, 36); $CF_2=CHF$ (87); $CF_2=CBrCl$ (30); $CF_2=CHCl$ (41); $CF_2=CFBr$ (75); $CF_2=NCF_3$ (82); $CF_2=CFCN$ (27); $CF_2=CFCOF$ (7); $CF_2=CFOCH_2CF_3$ (12); $CF_2=CBrCH_2N(CF_3)_2$ (30); $CF_2=CFCOCF_2CF_3$ (12); $CF_2=CHC_6H_5$ (33); $CF_2=CH(CH_2)_5CH_3$ (50); $CF_2=CH-Ar[Ar=aryl]$ (50)
Three Bonds		
Saturated (sp³)	0–16	CF_3CH_2F (16); CF_3CF_3 (3.5); CF_3CHF_2 (3); CH_2FCH_2F (10–12); $CF_2{}^aHCF^bHCF_2H$ ($J_{ab} = 13$); $CF_2HCF_2{}^aCH_2F$ ($J_{ab} = 14$); $CF_2{}^aCF_2{}^bCF^cHCH_3$ ($J_{ab} < l$; $J_{bc} = 15$); $CF_3{}^aCF^bHCF_2{}^cH$ ($J_{ab} = 12$; $J_{bc} = 12$); $CF_3{}^aCF_2{}^bC≡CCF_3$ ($J_{ab} = 3.3$); $CF_3{}^aCF_2{}^bC≡CCF_3$ ($J_{ab} = 3.3$; $(CF_3{}^a)_2CF^bC≡CCl$ ($J_{ab} = 10$); $CF_3CF_2COCH_2CH_3$ (1); $FCH_2CFHCO_2C_2H_5$ (−11.6); $CF_3{}^aCF_2{}^bCF^c COOH$ ($J_{ab} < l$; $J_{bc} < l$); $(CF_3{}^a)_2CF^bS(O)OC_2H_5$ ($J_{ab} = 8$)
Unsaturated (sp²)	>30	$FCH=CHF$ [cis (−18.7); trans (−133.5)]; $CF_2=CHBr$ (34.5); $CF_2=CHCl$ (41); $CF_2=CH_2$ (37)

^{13}C–^{19}F Coupling Constants

Fluorinated Family	J_{CF}, Hz	Examples
		One Bond
Alkanes	150–290	CH_3F (158); CH_2F_2 (237); CHF_3 (274); CF_4 (257); CF_3CF_3 (281); CF_3CH_3 (271); $(CH_3)_3CF$ (167); $(C^aF_3{}^b)_3C^cF_2{}^d$ [J_{ab} = 285; J_{cd} = 265]
Alkenes	250–300	$CF_2=CD_2$ (287); $CF_2=CCl_2$ (−289); $CF_2=CBr_2$ (290); $ClFC=CHCl$ [cis (−300); trans (−307)]; $ClFC=CClF$ [cis (290); trans (290)]
Alkynes	250–260	$C^aF_3{}^bC≡CF$ [J_{ab} = 259]; $CF_3C≡CCF_3$ (256)
Alkyl chlorides	275–350	$CFCl_3$ (337); CF_2Cl_2 (325); CF_3Cl (299); $CF_3(CCl_2)_2CF_3$ (286); CF_3CH_2Cl (274); $CF_3CCl=CCl_2$ (274); $CF_2=CCl_2$ (−289); CF_3CCl_3 (283)
Alkyl bromides	290–375	$CFBr_3$ (372); CF_2Br_2 (358); CF_3Br (324); CF_3CH_2Br (272); $CF_2=CBr_2$ (290)
Acyl fluorides	350–370	$HCOF$ (369); CH_3COF (353)
Carboxylic acids	245–290	CF_3COOH (283); CF_2HCO_2H (247)
Alcohols	≈275	CF_3CH_2OH (278)
Nitriles	≈250	CF_2HCN (244)
Esters	≈285	$CF_3CO_2CH_2CH_3$ (284)
Ketones	≈290	CF_3COCH_3 (289)
Ethers	≈265	$(CF_3)_2O$ (265)

^{31}P NMR ABSORPTIONS OF REPRESENTATIVE COMPOUNDS

^{31}P is considered to be a medium-sensitivity nucleus that has the advantage of yielding sharp lines over a very wide chemical-shift range. Its sensitivity is much less than that of ^{1}H, but it is superior to that of ^{13}C.[1]

The following charts provide information on the characteristic values for the ^{31}P spectra of representative phosphorus-containing compounds. The list is far from complete but gives an insight on the spectra of both organic and inorganic compounds. All data are presented in the form of a correlation chart. The reference in each case is 85% (mass/mass) phosphoric acid. The first chart provides the general chemical-shift range of the various phosphorus families and is followed by more-detailed charts that provide representative compounds for each of the families. These families are classified according to the coordination number around phosphorus, which ranges from 2 to 6.

This section only gives general information on ^{31}P NMR spectroscopy. The reader is advised to consult references 2 and 3, which include a large, detailed amount of updated spectral information and numerous references.

REFERENCES

1. NMR Lab, Department of Organic Chemistry, Hebrew University of Jerusalem; available online at http://drx.ch.huji.ac.il/nmr/; accessed 5/18/05.
2. Quin, L.D. and Verkade, J.G., Eds., *Phosphorus-31 NMR Spectral Properties in Compounds: Characterization and Structural Analysis,* John Wiley and Sons, New York, 1994.
3. Tebby, J.C., Ed., *Handbook of Phosphorus-31 Nuclear Magnetic Resonance Data,* CRC Press, Boca Raton, FL, 1991.

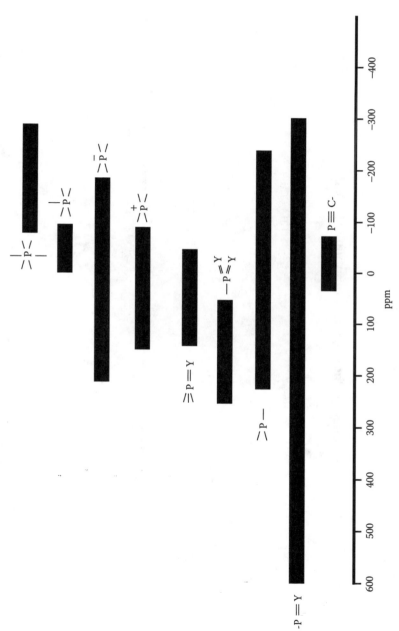

³¹P Chemical Shift Ranges for Various Phosphorus Compounds

Figure 3.37

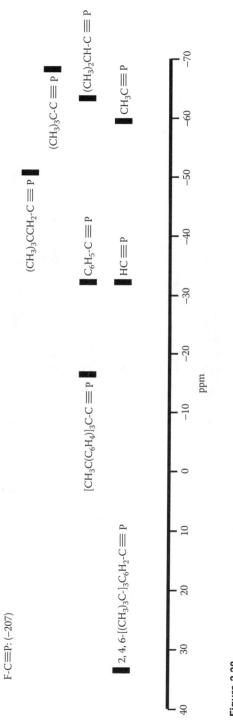

Figure 3.38

Dicoordinated Phosphorus Compounds (X=P−)

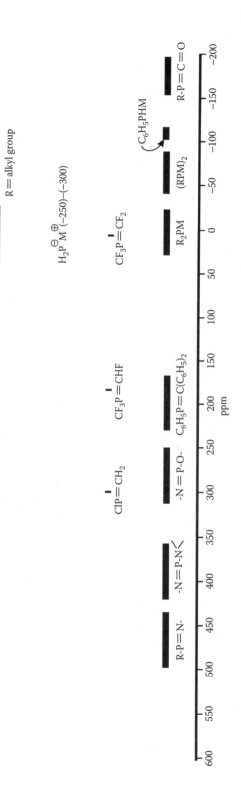

Figure 3.39

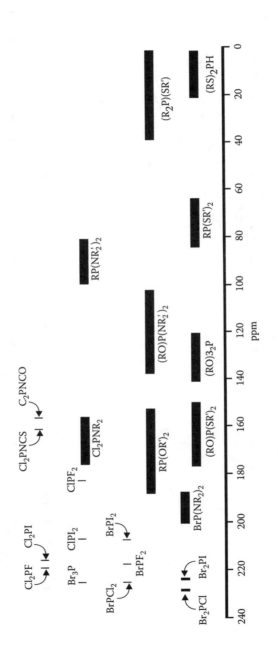

Figure 3.40

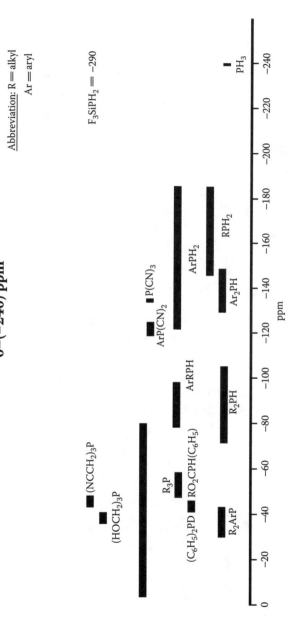

Tricoordinated Phosphorus Compounds ($\geqslant$P)
0–(−240) ppm

Abbreviation: R = alkyl
Ar = aryl

Figure 3.41

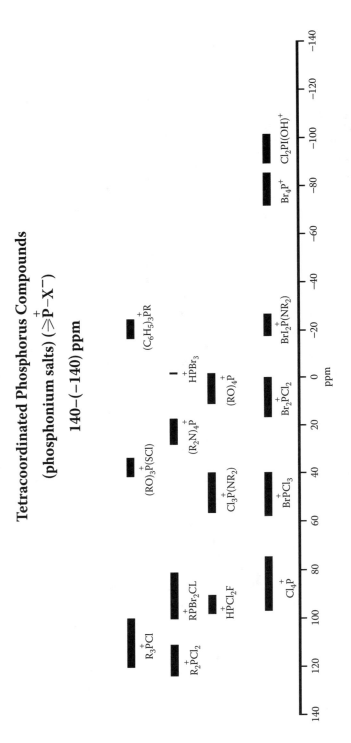

Tetracoordinated Phosphorus Compounds
(phosphonium salts) ($\geqslant\overset{+}{P}-X^-$)
140–(–140) ppm

Figure 3.42

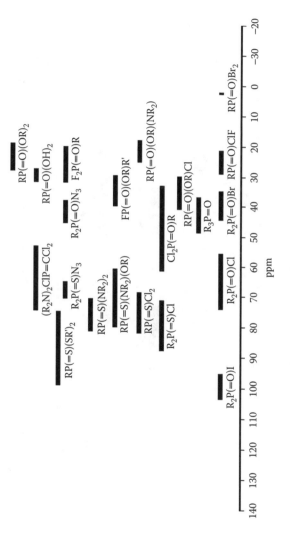

Figure 3.43

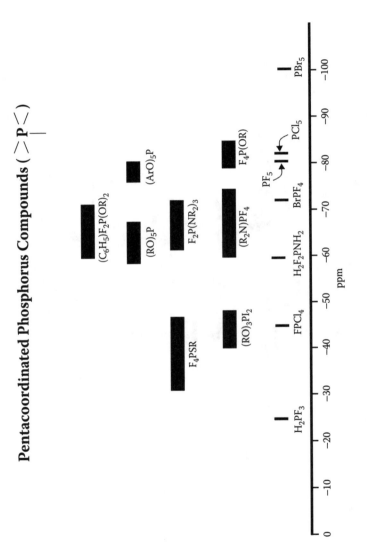

Figure 3.44

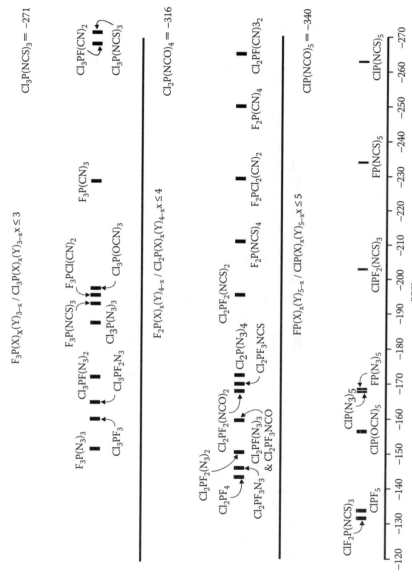

Figure 3.45

Hexacoordinated Phosphorus Compounds

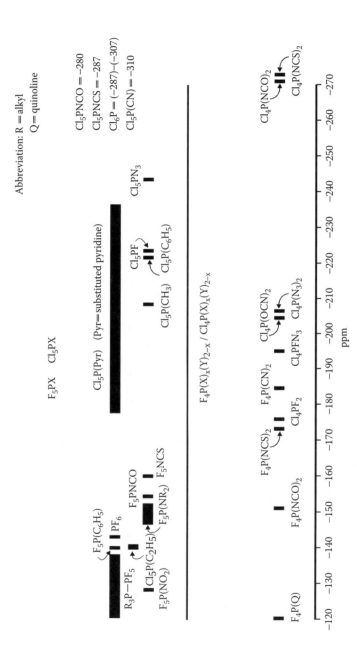

Figure 3.46

^{29}Si NMR ABSORPTIONS OF MAJOR CHEMICAL FAMILIES

The following correlation tables provide the regions of ^{29}Si nuclear magnetic resonance absorptions of some major chemical families. These absorptions are reported in the dimensionless units of parts per million (ppm) versus the standard compound tetramethylsilane (TMS, $(CH_3)_4Si$), which is recorded as 0.0 ppm.

^{29}Si NMR (natural abundance 4.7%) is a low-sensitivity nucleus that has a wide chemical-shift range that is useful in determining the identity of certain silicon-containing compounds, many of which are important in biological systems. It should be noted that when taking such spectra, there is always a background signal that is attributed to the glass composing the measurement tube. Modifying the probe, which can be costly, or a simple adjustment in the pulse sequence often overcomes the background signals. The ^{29}Si sensitivity is approximately 7.85×10^{-2} (at constant field) and 0.199 (at constant frequency) of that of ^{1}H.

For more detail concerning the chemical shifts, the reader is referred to the general references below and the literature cited therein.[1–7]

REFERENCES

1. Williams, E.A. and Cargioli, J.D., Silicon-29 NMR spectroscopy, *Ann. Rep. NMR Spectr.*, 9, 221, 1979; 15, 235, 1983.
2. Schraml, J. and Bellama J.M., ^{29}Si nuclear magnetic resonance, in *Determination of Organic Structures by Physical Methods,* Vol. 6, Nachod, F.C. and Zuckerman, J.J., Eds., Academic Press, New York, 1976.
3. Williams, E.A., NMR spectroscopy of organosilicon compounds, in *Chemistry and Physics of DNA-Ligand Interactions,* Kallenboch, N.R., Ed., Adenine Press, New York, 1990.
4. Schraml, J., ^{29}Si NMR spectroscopy of trimethyl silyl tags, in *Progress in NMR Spectroscopy,* Emsley, J.W., Feeneym J., and Sutcliffe, L.H., Eds., 22, 289, 1990.
5. Harris, R.K., in *Encyclopedia of Nuclear Magnetic Resonance,* Vol. 5, Granty, D.M. and Harris, R.K., Eds., John Wiley and Sons, Chichester, U.K., 1996.
6. Mason, J., *Multinuclear NMR,* Plenum Press, New York, 1987.
7. Some Useful NMR Data for Silicon Compounds, Department of Chemistry, Iowa State University; available on-line at http://avogadro.chem.iastate.edu/CHEM572/subpages/silicon.html; accessed 5/18/05.

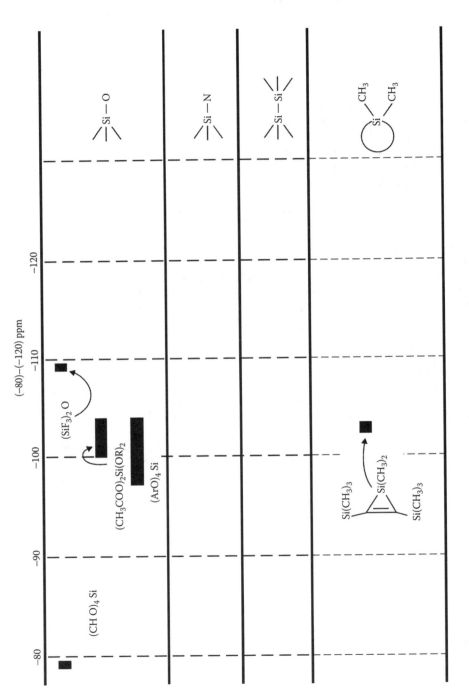

Figure 3.47

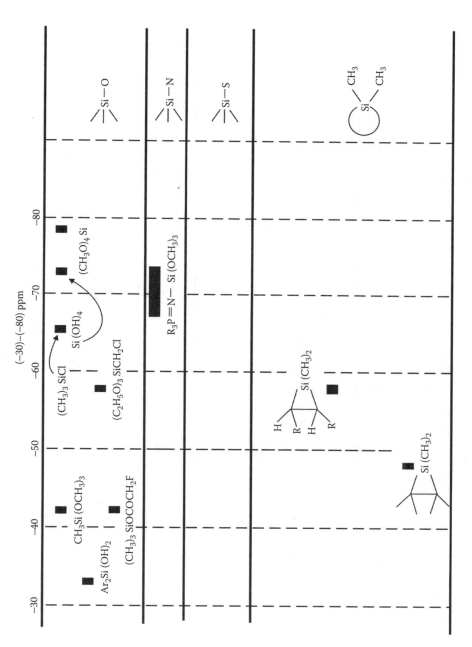

Figure 3.48

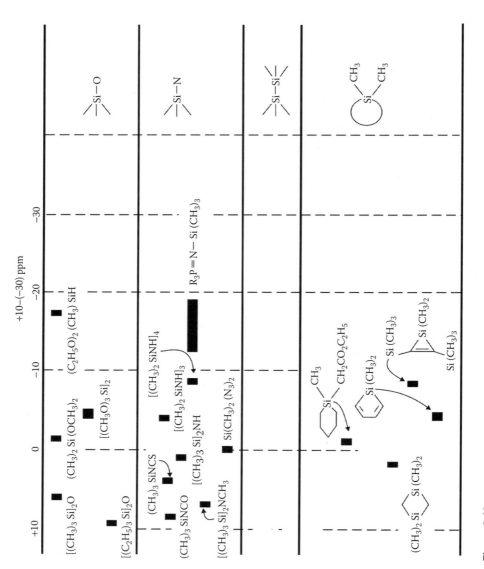

Figure 3.49

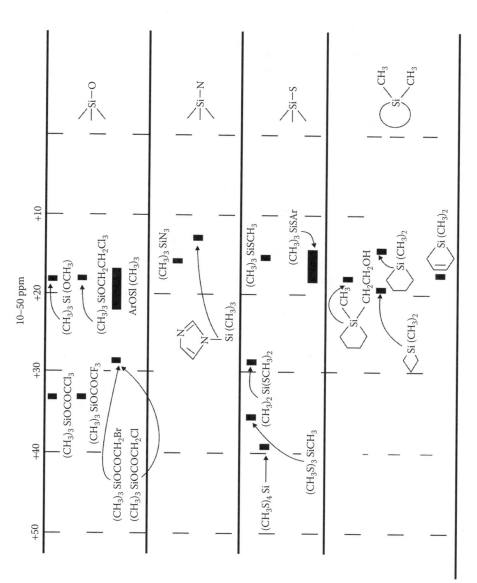

Figure 3.50

^{119}Sn NMR ABSORPTIONS OF MAJOR CHEMICAL FAMILIES

The following correlation tables provide the regions of ^{119}Sn nuclear magnetic resonance absorptions of some major chemical families. These absorptions are reported in the dimensionless units of parts per million (ppm) versus the standard compound tetramethylstanate $(CH_3)_4Sn$, which is recorded as 0.0 ppm.

^{119}Sn NMR is a low-sensitivity nucleus that has a relatively wide chemical-shift range, which is useful in determining the identification of certain tin-containing compounds. The ^{119}Sn sensitivity is approximately 5.18×10^{-2} (at constant field) that of ^{1}H. Its abundance (8.59%) is slightly higher than that of ^{117}Sn (7.68%) and much higher than that of ^{115}Sn (0.34%).

For more detail concerning the chemical shifts, the reader is referred to the general references below and the literature cited therein.[1–9] The reader should be aware that there is a great deal of chemical-shift variation when tin compounds are measured in different solvents. Moreover, many tin compounds are difficult to dissolve in common solvents.

REFERENCES

1. Petrosyan, V.S., NMR spectra and structures of organotin compounds, *Progr. NMR Spectroscopy,* 11, 115, 1978.
2. Smith, P.J., Chemical shifts of Sn-119 nuclei in organotin compounds, *Ann. Rep. NMR Spectroscopy,* 8, 292, 1978.
3. Wrackmeyer, B., Tin-119 NMR parameters, *Ann. Rep. NMR Spectroscopy,* 16, 73, 1985.
4. Hari, R. and Geanangel, R.A., Tin-119 NMR in coordination chemistry, *Coord. Chem. Rev.,* 44, 229, 1982.
5. Wrackmeyer, B., Multinuclear NMR and tin chemistry, *Chemistry in Britain,* 26, 48, 1990.
6. Kaur, A. and Sandhu, G.K., Use of ^{119}Sn Mossbauer and ^{119}Sn NMR spectroscopies in the study of organotin complexes, *J. Chem. Sci.,* 2, 1, 1986.
7. Harris, R.K., in *Encyclopedia of Nuclear Magnetic Resonance,* Vol. 5, Granty, D.M. and Harris, R.K., Eds., John Wiley and Sons, Chichester, U.K., 1996.
8. Mason, J., *Multinuclear NMR,* Plenum Press, New York, 1987.
9. Tin: NMR Data, WebElements; available on-line at http://www.webelements.com/webelements/elements/text/Sn/nucl.html; accessed 5/18/05.

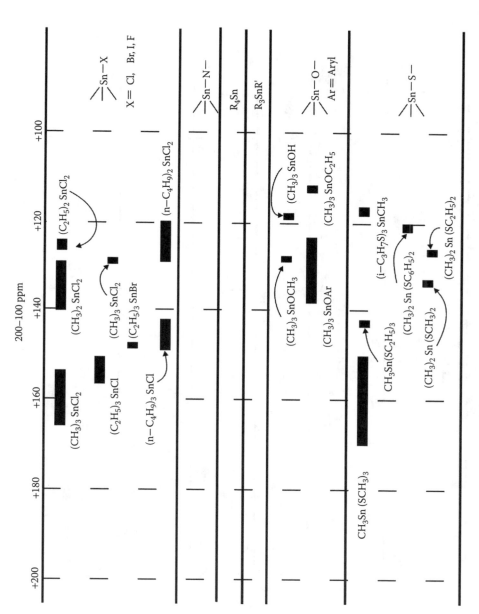

Figure 3.51

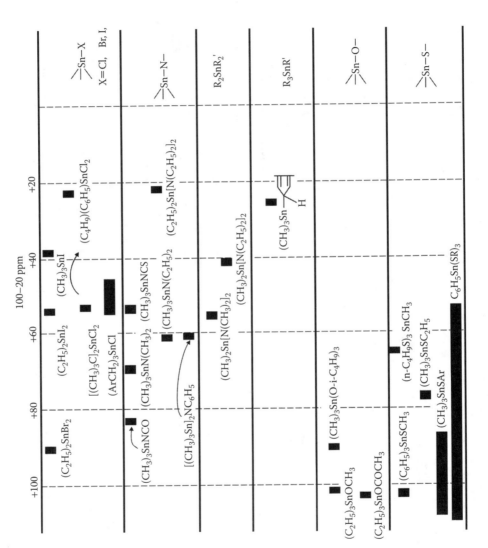

Figure 3.52

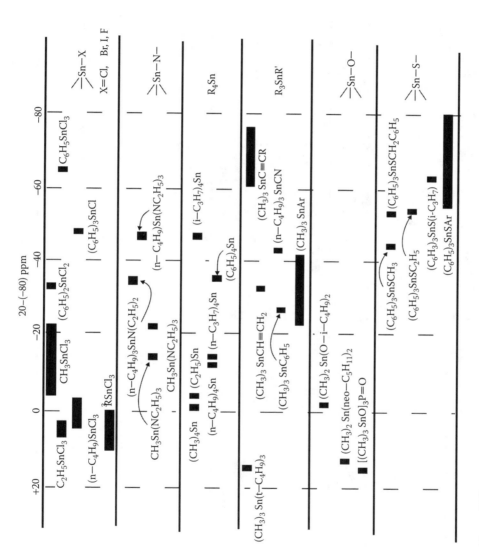

Figure 3.53

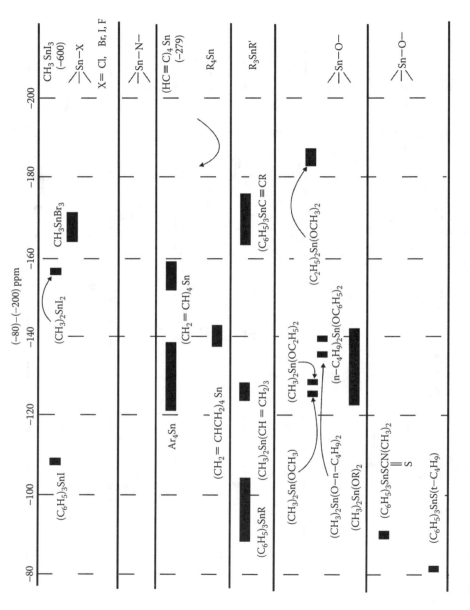

Figure 3.54

CHAPTER 4

Mass Spectrometry

CONTENTS

NATURAL ABUNDANCE OF IMPORTANT ISOTOPES

The following table lists the atomic masses and relative percent concentrations of naturally occurring isotopes of importance in mass spectrometry.[1–6]

REFERENCES

1. deHoffmann, E. and Stroobant, V., *Mass Spectrometry: Principles and Applications,* 2nd ed., John Wiley and Sons, Chichester, U.K., 2001.
2. Johnstone, R.A.W. and Rose, M.E., *Mass Spectrometry for Chemists and Biochemists,* Cambridge University Press, Cambridge, 1996.
3. Lide, D.R., Ed., *CRC Handbook of Chemistry and Physics,* 83rd ed., CRC Press, Boca Raton, FL, 2003.
4. McLafferty, F.W. and Turecek, F., *Interpretation of Mass Spectra,* 4th ed., University Science Books, Mill Valley, CA, 1993.
5. Watson, J.T., *Introduction to Mass Spectrometry,* 3rd ed., Lippincott-Raven, Philadelphia, 1997.
6. Bruno, T.J. and Svoronos, P.D.N., *Handbook of Basic Tables for Chemical Analysis,* 2nd ed., CRC Press, Boca Raton, FL, 2003.

Natural Abundance of Important Isotopes

Element	Total No. of Isotopes	Prominent Isotopes (mass, percent abundance)
Hydrogen	3	^{1}H (1.00783, 99.985) ^{2}H (2.01410, 0.015)
Boron	6	^{10}B (10.01294, 19.8) ^{11}B (11.00931, 80.2)
Carbon	7	^{12}C (12.00000, 98.9) ^{13}C (13.00335, 1.1)
Nitrogen	7	^{14}N (14.00307, 99.6) ^{15}N (15.00011, 0.4)
Oxygen	8	^{16}O (15.99491, 99.8) ^{18}O (17.9992, 0.2)
Fluorine	6	^{19}F (18.99840, ≈100.0)
Silicon	8	^{28}Si (27.97693, 92.2) ^{29}Si (28.97649, 4.7) ^{30}Si (29.97376, 3.1)
Phosphorus	7	^{31}P (30.97376, ≈100.0)
Sulfur	10	^{32}S (31.972017, 95.0) ^{33}S (32.97146, 0.7) ^{34}S (33.96786, 4.2)
Chlorine	11	^{35}Cl (34.96885, 75.5) ^{37}Cl (36.96590, 24.5)
Bromine	17	^{79}Br (78.9183, 50.5) ^{81}Br (80.91642, 49.5)
Iodine	23	^{127}I (126.90466, ≈100.0)

RULES FOR DETERMINATION OF MOLECULAR FORMULA

The following rules are used in the mass spectrometric determination of the molecular formula of an organic compound.[1-6] These rules should be applied to the molecular ion peak and its isotopic cluster. The molecular ion, in turn, is usually the highest mass in the spectrum. It must be an odd-electron ion, and must be capable of yielding all other important ions of the spectrum via a logical neutral-species loss. The elements that are assumed to be possibly present on the original molecule are carbon, hydrogen, nitrogen, the halogens, sulfur, and oxygen. The molecular formula that can be derived is not the only possible one, and consequently additional information from nuclear magnetic resonance spectrometry and infrared spectrophotometry is necessary for the final determination of the molecular formula.

Modern mass spectral databases allow the automated searching of very extensive mass spectral libraries. This has made the identification of compounds by mass spectrometry a far more straightforward task. One must understand, however, that such databases are no substitute for the careful analysis of each mass spectrum, and that the results of database matchup are merely suggestions.

REFERENCES

1. Lee, T.A., *A Beginner's Guide to Mass Spectral Interpretation,* Wiley, New York, 1998.
2. McLafferty, F.W., *Interpretation of Mass Spectra,* University Science Books, Mill Valley, CA, 1993.
3. Shrader, S.R., *Introductory Mass Spectrometry,* Allyn and Bacon, Boston, 1971.
4. Smith, R.M., *Understanding Mass Spectra: A Basic Approach,* Wiley, New York, 1999.
5. Watson, J.T. and Watson, T.J., *Introduction to Mass Spectrometry,* Lippincott, Williams and Wilkins, Philadelphia, 1998.
6. NIST, NIST Standard Reference Database 1A, NIST/EPA/NIH Mass Spectral Library with Search Program, Data Version NIST '02, Software Version 2.0.

Rule 1: An odd molecular ion value suggests the presence of an odd number of nitrogen atoms; an even molecular ion value is due to the presence of zero or an even number of nitrogen atoms. Thus, m/e = 141 suggests 1, 3, 5, 7, etc. nitrogen atoms, while m/e = 142 suggests 0, 2, 4, 6, etc., nitrogen atoms.

Rule 2: The maximum number of carbons (N_C^{max}) can be calculated from the formula

$$N_C^{max} = \frac{\text{Relative intensity of M}+1\text{ peak}}{\text{Relative intensity of M}^+\text{ peak}} \times \frac{100}{1.1}$$

where M + 1 is the peak one unit above the value of the molecular ion (M$^+$). This rule gives the *maximum* number of carbons, but not necessarily the *actual* number. If, for example, the relative intensities of M$^+$ and M + 1 are 100% and 9%, respectively, then the maximum number of carbons is

$$(N_C^{max}) = (9/100) \times (100/1.1) = 8$$

In this case there is a possibility for seven, six, etc., carbons, but not for nine or more.

Rule 3: The maximum numbers of sulfur atoms (N_S^{max}) can be calculated from the formula

$$N_S^{max} = \frac{\text{Relative intensity of M}+2\text{ peak}}{\text{Relative intensity of M}^+\text{ peak}} \times \frac{100}{4.4}$$

where M + 2 is the peak two units above that of the molecular ion M$^+$.

Rule 4: The actual number of chlorine or bromine atoms can be derived from the table in the section entitled "Chlorine–Bromine Combination Isotope Intensities."

Rule 5: The difference from the M^+ value should be attributed only to oxygen and hydrogen atoms. These rules assume the absence of phosphorus, silicon, or any other elements, including sulfur.

NEUTRAL MOIETIES EJECTED FROM SUBSTITUTED
BENZENE RING COMPOUNDS

The following table lists the most common substituents encountered in benzene rings and the neutral particles lost and observed on the mass spectrum. Complex rearrangements are often encountered and are enhanced by the presence of one or more heteroatomic substituent(s) in the aromatic compound. All neutral particles that are not the product of rearrangement appear in parentheses and are produced alongside the species that are formed via rearrangement. Prediction of the more abundant moiety is not easy, as it is seriously affected by factors that dictate the nature of the compound. These include the nature and the position of any other substituents as well as the stability of any intermediate(s) formed. Correlations of the data with the corresponding Hammett σ constants have been neither consistent nor conclusive.

REFERENCE

1. Rose, M.E. and Johnstone, R.A.W., *Mass Spectroscopy for Chemists and Biochemists,* Cambridge University Press, Cambridge, 1982.

Neutral Moieties Ejected from Substituted Benzene Ring Compounds

Substituent	Neutral Moiety(s) Ejected after Rearrangement
NO_2	NO, CO, (NO_2)
NH_2	HCN
$NHCOCH_3$	C_2H_2O, HCN
CN	HCN
F	C_2H_2
OCH_3	CH_2O, CHO, CH_3
OH	CO, CHO
SO_2, NH_2	SO_2, HCN
SH	CS, CHS, (SH)
SCH_3	CS, CH_2S, SH, (CH_3)

ORDER OF FRAGMENTATION INITIATED BY THE PRESENCE OF A SUBSTITUENT ON A BENZENE RING

The following table lists the relative order of ease of fragmentation that is initiated by the presence of a substituent in the benzene ring in mass spectroscopy. The ease of fragmentation decreases from top to bottom. The substituents marked with an asterisk (*) are very similar in their ease of fragmentation. Particularly in the case of disubstituted benzene rings, the order of fragmentation at the substituent linkage may be easily predicted using this table. As a rule of thumb, the more complex the size of the substituent, the easier its decomposition. For instance, in all chloroace-tophenone isomers (1,2-, 1,3-, or 1,4-), the elimination of the methyl radical occurs before the loss of chlorine. On the other hand, under normal mass conditions, all bromofluorobenzenes (1,2-, 1,3-, and 1,4-) easily lose the bromine but not the fluorine. Deuterium-labeling studies have indicated that any rearrangement of the benzene compounds occurs in the molecular ion and before fragmentation.

REFERENCE

1. Rose, M.E. and Johnstone, R.A.W., *Mass Spectroscopy for Chemists and Biochemists,* Cambridge University Press, Cambridge, 1982.

Order of Fragmentation Initiated by the Presence of a Substituent on a Benzene Ring

Substituent	Neutral Moiety Eliminated
$COCH_3$	CH_3
CO_2CH_3	OCH_3
NO_2	NO_2
* I	I
* OCH_3	CH_2O, CHO
* Br	Br
OH	CO, CHO
CH_3	H
Cl	Cl
NH_2	HCN
CN	HCN
F	C_2H_2

CHLORINE–BROMINE COMBINATION ISOTOPE INTENSITIES

Due to the distinctive mass spectral patterns caused by the presence of chlorine and bromine in a molecule, interpretation of a mass spectrum can be much easier if the results of the relative isotopic concentrations are known. The following table provides peak intensities (relative to the molecular ion (M^+) at an intensity normalized to 100%) for various combinations of chlorine and bromine atoms, assuming the absence of all other elements except carbon and hydrogen.[1-5] The mass-abundance calculations were based upon the most recent atomic mass data.[1]

REFERENCES

1. Lide, D.R., *CRC Handbook of Chemistry and Physics,* 83rd ed., CRC Press, Boca Raton, FL, 2003.
2. McLafferty, F.W., *Interpretation of Mass Spectra,* 4th ed., University Science Books, Mill Valley, CA, 1993.
3. Silverstein, R.H., Bassler, G.C., and Morrill, T.C., *Spectroscopic Identification of Organic Compounds,* 6th ed., John Wiley and Sons, New York, 1998.
4. Williams, D.H. and Fleming, I., *Spectroscopic Methods in Organic Chemistry,* 4th ed., McGraw-Hill, London, 1989.
5. Bruno, T.J. and Svoronos, P.D.N., *Handbook of Basic Tables for Chemical Analysis,* 2nd ed., CRC Press, Boca Raton, FL, 2003.

Relative Intensities of Isotope Peaks for Combinations of Bromine and Chlorine ($M^+ = 100\%$)

		Br_0	Br_1	Br_2	Br_3	Br_4
Cl_0	P + 2		98.0	196.0	294.0	390.8
	P + 4			96.1	288.2	574.7
	P + 6				94.1	375.3
	P + 8					92.0
Cl_1	P + 2	32.5	130.6	228.0	326.1	424.6
	P + 4		31.9	159.0	383.1	704.2
	P + 6			31.2	187.4	564.1
	P + 8				30.7	214.8
	P + 10					30.3
Cl_2	P + 2	65.0	163.0	261.1	359.3	456.3
	P + 4	10.6	74.4	234.2	490.2	840.3
	P + 6		10.4	83.3	312.8	791.6
	P + 8			10.2	91.7	397.5
	P + 10				9.8	99.2
	P + 12					10.1
Cl_3	P + 2	97.5	195.3	294.0	393.3	489
	P + 4	31.7	127.0	99.7	609.8	989
	P + 6	3.4	34.4	159.4	473.8	1064
	P + 8		3.3	37.1	193.9	654
	P + 10			3.2	39.6	229
	P + 12				3.0	42
	P + 14					3.2
Cl_4	P + 2	130.0	228.3	326.6	4.2	522
	P + 4	63.3	190.9	414.9	735.3	1149
	P + 6	13.7	75.8	263.1	670.0	1388
	P + 8	1.2	14.4	88.8	347.1	1002
	P + 10		1.1	15.4	102.2	443
	P + 12			1.3	16.2	117
	P + 14				0.7	17
Cl_5	P + 2	162.6	260.7	358.9		
	P + 4	105.7	265.3	520.8		
	P + 6	34.3	137.9	397.9		
	P + 8	5.5	39.3	174.5		
	P + 10	0.3	5.8	44.3		
	P + 12		0.3	5.7		
	P + 14			0.5		
Cl_6	P + 2	195.3				
	P + 4	158.6				
	P + 6	68.8				
	P + 8	16.6				
	P + 10	2.1				
	P + 12	0.1				
Cl_7	P + 2	227.8				
	P + 4	222.1				
	P + 6	120.3				
	P + 8	39.0				
	P + 10	7.5				
	P + 12	0.8				
	P + 14	0.05				

REFERENCE COMPOUNDS UNDER ELECTRON-IMPACT CONDITIONS IN MASS SPECTROMETRY

The following table lists the most popular reference compounds for use under electron-impact conditions in mass spectrometry. For accurate mass measurements, the reference compound is introduced and ionized concurrently with the sample, and the reference peaks are resolved from sample peaks. Reference compounds should contain as few heteroatoms and isotopes as possible. This is to facilitate the assignment of reference masses and minimize the occurrence of unresolved multiplets within the reference spectrum.[1] An approximate upper mass limit should assist in the selection of the appropriate reference.[1,2]

REFERENCES

1. Chapman, J.R., *Computers in Mass Spectrometry,* Academic Press, London, 1978.
2. Chapman, J.R., *Practical Organic Mass Spectrometry,* 2nd ed., John Wiley and Sons, Chichester, U.K., 1995.

Reference Compounds under Electron-Impact Conditions in Mass Spectrometry

Reference Compound	Formula	Upper Mass Limit
perfluoro-2-butyltetrahydrofuran	$C_8F_{16}O$	416
decafluorotriphenyl phosphine (ultramark 443; DFTPP)	$(C_6F_5)_3P$	443
heptacosafluorotributylamine (perfluoro tributylamine; heptacosa; PFTBA)	$(C_4F_9)_3N$	671
perfluoro kerosene, low-boiling (perfluoro kerosene-L)	$CF_3(CF_2)_nCF_3$	600
perfluoro kerosene, high-boiling (perfluoro kerosene-H)	$CF_3(CF_2)_nCF_3$	800–900
Tris (trifluoromethyl)-s-triazine	$C_3N_3(CF_3)_3$	285
Tris (pentafluoroethyl)-s-triazine	$C_3N_3(CF_2CF_3)_3$	435
Tris (heptafluoropropyl)-s-triazine	$C_3N_3(CF_2CF_2CF_3)_3$	585
Tris (perfluoroheptyl)-s-triazine	$C_3N_3[(CF_2)_6CF_3]_3$	1185
Tris (perfluorononyl)-s-triazine	$C_3N_3[(CF_2)_8CF_3]_3$	1485
Ultramark 1621 (fluoroalkoxy cyclotriphosphazine mixture)	$P_3N_3[OCH_2(CF_2)_nH]_6$	≈2000
Fomblin diffusion pump fluid (Ultramark F-series; perfluoropolyether)	$CF_3O[CF(CF_3)CF_2O]_m(CF_2O)_nCF_3$	≥3000

MAJOR REFERENCE MASSES IN THE SPECTRUM OF HEPTACOSAFLUOROTRIBUTYLAMINE (PERFLUOROTRIBUTYLAMINE)

The following list tabulates the major reference masses (with their relative intensities and formulas) of the mass spectrum of heptacosafluorotributylamine.[1] This is one of the most widely used reference compounds in mass spectrometry.

REFERENCE

1. Chapman, J.R., *Practical Organic Mass Spectrometry*, 2nd ed., John Wiley and Sons, Chichester, U.K., 1995.

Major Reference Masses in the Spectrum of Heptacosafluorotributylamine (Perfluorotributylamine)

Mass	Relative Intensity	Formula	Mass	Relative Intensity	Formula
613.9647	2.6	$C_{12}F_{24}N$	180.9888	1.9	C_4F_7
575.9679	1.7	$C_{12}F_{22}N$	175.9935	1.0	C_4F_6N
537.9711	0.4	$C_{12}F_{20}N$	168.9888	3.6	C_3F_7
501.9711	8.6	$C_9F_{20}N$	163.9935	0.7	C_3F_6N
463.9743	3.8	$C_9F_{18}N$	161.9904	0.3	C_4F_6
425.9775	2.5	$C_9F_{16}N$	149.9904	2.1	C_3F_6
413.9775	5.1	$C_8F_{16}N$	130.9920	31	C_3F_5
375.9807	0.9	$C_8F_{14}N$	118.9920	8.3	C_2F_5
325.9839	0.4	$C_7F_{12}N$	113.9967	3.7	C_2F_4N
313.9839	0.4	$C_6F_{12}N$	111.9936	0.7	C_3F_4
263.9871	10	$C_5F_{10}N$	99.9936	12	C_2F_4
230.9856	0.9	C_5F_9	92.9952	1.1	C_3F_3
225.9903	0.6	C_5F_8N	68.9952	100	CF_3
218.9856	62	C_4F_9	49.9968	1.0	CF_2
213.9903	0.6	C_4F_8N	30.9984	2.3	CF

COMMON FRAGMENTATION PATTERNS OF FAMILIES OF ORGANIC COMPOUNDS

The following table provides a guide to the identification and interpretation of commonly observed mass spectral fragmentation patterns for common organic functional groups.[1–10] It is of course highly desirable to augment mass spectroscopic data with as much other structural information as possible. Especially useful in this regard will be the confirmatory information of infrared and ultraviolet spectrophotometry as well as nuclear magnetic resonance spectrometry.

REFERENCES

1. Bowie, J.H., Williams, D.H., Lawesson, S.O., Madsen, J.O., Nolde, C., and Schroll, G., Studies in mass spectrometry, XV: mass spectra of sulphoxides and sulphones: the formation of C–C and C–O bonds upon electron impact, *Tetrahedron*, 22, 3515, 1966.
2. Johnstone, R.A.W. and Rose, M.E., *Mass Spectrometry for Chemical and Biochemists,* Cambridge University Press, Cambridge, 1996.
3. Lee, T.A., *A Beginner's Guide to Mass Spectral Interpretation,* Wiley, New York, 1998.
4. McLafferty, F.W., *Interpretation of Mass Spectra,* 4th ed., University Science Books, Mill Valley, CA, 1993.
5. Pasto, D.J. and Johnson, C.R., *Organic Structure Determination,* Prentice-Hall, Englewood Cliffs, NJ, 1969.
6. Silverstein, R.M., Bassler, G.C., and Morrill, T.C., *Spectroscopic Identification of Organic Compounds,* 6th ed., John Wiley and Sons, New York, 1998.
7. Smakman, R. and deBoer, T.J., The mass spectra of some aliphatic and alicyclic sulphoxides and sulphones, *Org. Mass Spec.,* 3, 1561, 1970.
8. Smith, R.M., *Understanding Mass Spectra: A Basic Approach,* Wiley, New York, 1999.
9. Watson, T.J. and Watson, J.T., *Introduction to Mass Spectrometry,* Lippincott, Williams and Wilkins, Philadelphia, 1997.
10. Bruno, T.J. and Svoronos, P.D.N., *Handbook of Basic Tables for Chemical Analysis,* 2nd ed., CRC Press, Boca Raton, FL, 2003.

Common Fragmentation Patterns of Families of Organic Compounds

Family	Molecular Ion Peak	Common Fragments; Characteristic Peaks
Acetals	—	Cleavage of all C–O, C–H, and C–C bonds around the original aldehydic carbon
Alcohols	Weak for 1° and 2°; not detectable for 3°; strong for benzyl alcohols	Loss of 18 (H_2O, usually by cyclic mechanism); loss of H_2O and olefin simultaneously with four (or more) carbon-chain alcohols; prominent peak at m/e = 31 ($CH_2\overset{+}{O}H$)$^+$ for 1° alcohols; prominent peak at m/e = ($RCH\overset{..}{O}H$)$^+$ for 2° and m/e = ($R_2C\overset{..}{O}H$)$^+$ for 3° alcohols
Aldehydes	Low intensity	Loss of aldehydic hydrogen (strong M–1 peak, especially with aromatic aldehydes); strong peak at m/e = 29 ($HC{\equiv}O^+$); loss of chain attached to alpha carbon (beta cleavage); McLafferty rearrangement via beta cleavage if gamma hydrogen is present
Alkanes		
Chain	Low intensity	Loss of 14 units (CH_2)
Branched	Low intensity	Cleavage at the point of branch; low-intensity ions from random rearrangements
Alicyclic	Rather intense	Loss of 28 units ($CH_2{=}CH_2$) and side chains
Alkenes (olefins)	Rather high intensity (loss of π-electron), especially in the case of cyclic olefins	Loss of units of general formula C_nH_{2n-1}; formation of fragments of the composition C_nH_{2n} (via McLafferty rearrangement); retro Diels–Alder fragmentation
Alkyl halides	Abundance of molecular ion F < Cl < Br < I; intensity decreases with increase in size and branching	Loss of fragments equal to the mass of the halogen until all halogens are cleaved off
fluorides	Very low intensity Low intensity; characteristic isotope cluster	Loss of 20 (HF); loss of 26 (C_2H_2) in case of fluorobenzenes
chlorides	Low intensity; characteristic isotope cluster	Loss of 35 (Cl) or 36 (HCl); loss of chain attached to the gamma carbon to the carbon carrying the Cl
bromides	Higher than other halides	Loss of 79 (Br); loss of chain attached to the gamma carbon to the carbon carrying the Br
iodides		Loss of 127 (I)
Alkynes	Rather high intensity (loss of π-electron)	Fragmentation similar to that of alkenes
Amides	High intensity	Strong peak at m/e = 44 indicative of a 1° amide ($O{=}C{=}N^+H_2$); base peak at m/e = 59 ($H_2{=}C(OH)N^+H_2$); possibility of McLafferty rearrangement; loss of 42 (C_2H_2O) for amides of the form $RNHCOCH_3$ when R is aromatic ring
Amines	Hardly detectable in case of acyclic aliphatic amines; high intensity for aromatic and cyclic amines	Beta cleavage yielding >C=N$^+$<; base peak for all 1° amines at m/e = 30 ($CH_2{=}N^+H_2$); moderate M – 1 peak for aromatic amines; loss of 27 (HCN) in aromatic amines; fragmentation at alpha carbons in cyclic amines
Aromatic hydrocarbons (arenes)	Rather intense	Loss of side chain; formation of RCH=CHR′ (via McLafferty rearrangement); cleavage at the bonds beta to the aromatic ring; peaks at m/e = 77 (benzene ring, especially monosubstituted) and 91 (tropyllium); the ring position of alkyl substitution has very little effect on the spectrum
Carboxylic acids	Weak for straight-chain monocarboxylic acids; large if aromatic acids	Base peak at m/e = 60 ($CH_2{=}C(OH)_2$) if alpha-hydrogen is present; peak at m/e = 45 (COOH); loss of 17 (–OH) in case of aromatic acids or short-chain acids
Disulfides	Rather low intensity	Loss of olefins (m/e equal to R–S–S–H$^+$); strong peak at m/e = 66 (HSSH$^+$)
Phenols	Highly intense peak (base peak,[a] generally)	Loss of 28 (C=O) and 29 (CHO); strong peak at m/e = 65 ($C_5H_5^+$)
Sulfides (thioethers)	Rather low-intensity peak, but higher than that of corresponding ether	Similar to those of ethers (–O– substituted by –S–); aromatic sulfides show strong peaks at m/e = 109 ($C_6H_5S^+$), 65 ($C_5H_5^+$), 91 (tropyllium ion)

(continued)

Common Fragmentation Patterns of Families of Organic Compounds (Continued)

Family	Molecular Ion Peak	Common Fragments; Characteristic Peaks
Sulfonamides	Intense	Loss of m/e = 64 ($SONH_2$) and m/e = 27 (HCN) in the case of benzenesulfonamide
Esters	Weak	Base peak at m/e equal to the mass of $R-C≡O^+$; peaks at m/e equal to the mass of $^+O≡C-OR'$ and the mass of OR' and R'; McLafferty rearrangement possible in the case of (a) presence of a beta hydrogen in R' (peak at m/e equal to the mass of $R-C(^+OH)OH$) and (b) presence of a gamma hydrogen in R (peak at m/e equal to the mass of $CH_2=C(^+OH)OR$); loss of 42 ($CH_2=C=O$) in the case of benzyl esters; loss of ROH via the ortho effect in the case of o-substituted benzoates
Ketones	High-intensity peak	Loss of R-groups attached to the >C=O (alpha cleavage); peak at m/e = 43 for all methyl ketones (CH_3CO^+); McLafferty rearrangement via beta cleavage if gamma hydrogen is present; loss of m/e = 28 (C=O) for cyclic ketones after initial alpha cleavage and McLafferty rearrangement
Mercaptan (thiols)	Rather low intensity, but higher than that of corresponding alcohol	Similar to those of alcohols (−OH substituted by −SH); loss of m/e = 45 (CHS) and m/e = 44 (CS) for aromatic thiols
Nitriles	Unlikely to be detected except in the case of acetonitrile (CH_3CN) and propionitrile (C_2H_5CN)	M + 1 ion may appear (especially at higher pressures); M − 1 peak is weak but detectable ($R-CH=C=N^+$); base peak at m/e = 41 ($CH_2=C=N^+H$); McLafferty rearrangement possible; loss of HCN in the case of cyanobenzenes
Nitrites	Absent (or very weak at best)	Base peak at m/e = 30 (NO^+); large peak at m/e = 60 ($CH_2=O^+NO$) in all unbranched nitrites at the alpha carbon; absence of m/e = 46 allows differentiation from nitrocompounds
Nitro compounds	Seldom observed	Loss of 30 (NO); subsequent loss of CO (in the case of aromatic nitrocompounds); loss of NO_2 from molecular ion peak
Sulfones	High intensity	Similar to sulfoxides; loss of mass equal to RSO_2; aromatic heterocycles show peaks at M − 32 (sulfur), M − 48 (SO), M − 64 (SO_2)
Sulfoxides	High intensity	Loss of 17 (OH); loss of alkene (m/e equal to $RSOH^+$); peak at m/e = 63 ($CH_2=SOH)^+$; aromatic sulfoxides show peak at m/e = 125 ($C_6H_5SO^+$), 97 ($C_5H_5S^+$), 93 ($C_6H_5O^+$); aromatic heterocycles show peaks at M −16 (oxygen), M − 29 (COH), M − 48 (SO)

[a] The base peak is the most intense peak in the mass spectrum, and it is often the molecular ion peak, M^+.

COMMON FRAGMENTS LOST

The following table gives a list of neutral species that are most commonly lost when measuring the mass spectra of organic compounds. The list is suggestive rather than comprehensive, and it should be used in conjunction with other sources.[1-5] The listed fragments include only combinations of carbon, hydrogen, oxygen, nitrogen, sulfur, and the halogens.

REFERENCES

1. Hamming, M. and Foster, N., *Interpretation of Mass Spectra of Organic Compounds,* Academic Press, New York, 1972.
2. McLafferty, F.W., *Interpretation of Mass Spectra,* 4th ed., University Science Books, Mill Valley, CA, 1993.
3. Silverstein, R.M., Bassler, G.C., and Morrill, T.C., *Spectroscopic Identification of Organic Compounds,* 6th ed., John Wiley and Sons, New York, 1996.
4. Bruno, T.J., *CRC Handbook for the Analysis and Identification of Alternative Refrigerants,* CRC Press, Boca Raton, FL, 1995.
5. Bruno, T.J. and Svoronos, P.D.N., *Handbook of Basic Tables for Chemical Analysis*, 2nd ed., CRC Press, Boca Raton, FL, 2003.

Common Fragments Lost

Mass Lost	Fragment Lost	Mass Lost	Fragment Lost
1	H•	49	•CH_2Cl
15	CH_3•	51	•CHF_2
17	OH•	52	C_4H_4•, C_2N_2
18	H_2O	54	$CH_2=CHCH=CH_2$
19	F•	55	$CH_2=CH–CH$•CH_3
20	HF	56	$CH_2=CH–CH_2CH_3$, $CH_3CH=CHCH_3$,
26	HC≡CH, •C≡N		CO (2 moles)
27	$CH_2–CH$•, HC≡N	57	C_4H_9•
28	$CH_2=CH_2$, C=O, (HCN and H•)	58	•NCS, $(CH_3)_2$C=O, (NO and CO)
29	CH_3CH_2•, H–•C=O	59	CH_3OCO•, CH_3CONH_2, C_2H_3S•
30	•CH_2NH_2, HCHO, NO	60	C_3H_7OH
31	CH_3O•, •CH_2OH, CH_3NH_2	61	CH_3CH_2S•, $(CH_2)_2S$•H
32	CH_3OH, S	62	H_2S, and $CH_2=CH_2$
33	HS•	63	•CH_2CH_2Cl
34	H_2S	64	S_2•, SO_2•, C_5H_4•
35	Cl•	68	$CH_2=CHC(CH_3)=CH_2$
36	HCl, $2H_2O$	69	CF_3•, C_5H_9•
37	H_2Cl	71	C_5H_{11}•
38	C_3H_2•, C_2N, F_2	73	CH_3CH_2OC•=O
39	C_3H_3, HC_2N	74	C_4H_9OH
40	$CH_3C≡CH$	75	C_6H_3
41	$CH_2=CHCH_2$•	76	C_6H_4, CS_2
42	$CH_2=CHCH_3$, $CH_2=C=O$, $(CH_2)_3$, NCO,	77	C_6H_5•, HCS_2•
	$NCNH_2$	78	C_6H_6, H_2CS_2•, C_5H_4N
43	C_3H_7•, $CH_3C=O$•, $CH_2=CH–O$•, HCNO	79	Br•, C_5H_5N
44	$CH_2=CHOH$, CO_2, N_2O, $CONH_2$,	80	HBr
	$NHCH_2CH_3$	85	•$CClF_2$
45	CH_3CHOH, CH_3CH_2O•, CO_2H,	100	$CF_2=CF_2$
	$CH_3CH_2NH_2$	119	CF_3CF_2•
46	CH_3CH_2OH, •NO_2	122	$C_6H_5CO_2H$
47	CH_3S•	127	I•
48	CH_3SH, SO, O_3	128	HI

IMPORTANT PEAKS IN THE MASS SPECTRA OF COMMON SOLVENTS

The following table gives the most important peaks that appear in the mass spectra of the most common solvents that can be found as an impurity in organic samples. The solvents are classified in ascending order, based upon their M^+ peaks. The highest intensity peaks are indicated by 100%.[1-5]

REFERENCES

1. Clere, J.T., Pretsch, E., and Seibl, J., *Studies in Analytical Chemistry I. Structural Analysis of Organic Compounds by Combined Application of Spectroscopic Methods,* Elsevier, Amsterdam, 1981.
2. McLafferty, F.W., *Interpretation of Mass Spectra,* 4th ed., University Science Books, Mill Valley, CA, 1993.
3. Pasto, D.J. and Johnson, C.R., *Organic Structure Determination,* Prentice Hall, Englewood Cliffs, NJ, 1969.
4. Smith, R.M., *Understanding Mass Spectra: A Basic Approach,* Wiley, New York, 1999.
5. Bruno, T.J. and Svoronos, P.D.N., *Handbook of Basic Tables for Chemical Analysis,* 2nd ed., CRC Press, Boca Raton, FL, 2003.

Important Peaks in the Mass Spectra of Common Solvents

Solvents	Formula	M^+	Important Peaks (m/e)
water	H_2O	18 (100%)	17
methanol	CH_3OH	32	31 (100%), 29, 15
acetonitrile	CH_3CN	41 (100%)	40, 39, 38, 28, 15
ethanol	CH_3CH_2OH	46	45, 31 (100%), 27, 15
dimethylether	CH_3OCH_3	46 (100%)	45, 29, 15
acetone	CH_3COCH_3	58	43 (100%), 42, 39, 27, 15
acetic acid	CH_3CO_2H	60	45, 43, 18, 15
ethylene glycol	$HOCH_2CH_2OH$	62	43, 33, 31 (100%), 29, 18, 15
furan	C_4H_4O	68 (100%)	42, 39, 38, 31, 29, 18
tetrahydrofuran	C_4H_8O	72	71, 43, 42 (100%), 41, 40, 39, 27, 18, 15
n-Pentane	C_5H_{12}	72	57, 43 (100%), 42, 41, 39, 29, 28, 27, 15
dimethylformamide (DMF)	$HCON(CH_3)_2$	73 (100%)	58, 44, 42, 30, 29, 28, 18, 15
diethylether	$(C_2H_5)_2O$	74	59, 45, 41, 31 (100%), 29, 27, 15
methylacetate	$CH_3CO_2CH_3$	74	59, 43 (100%), 42, 32, 29, 28, 15
carbon disulfide	CS_2	76 (100%)	64, 44, 32
benzene	C_6H_6	78 (100%)	77, 52, 51, 50, 39, 28
pyridine	C_5H_5N	79 (100%)	80, 78, 53, 52, 51, 50, 39, 26
dichloromethane	CH_2Cl_2	84	86, 51, 49 (100%), 48, 47, 35, 28
cyclohexane	C_6H_{12}	84	69, 56, 55, 43, 42, 41, 39, 27
n-Hexane	C_6H_{14}	86	85, 71, 69, 57 (100%), 43, 42, 41, 39, 29, 28, 27
p-Dioxane	$C_4H_8O_2$	88 (100%)	87, 58, 57, 45, 43, 31, 30, 29, 28
tetramethylsilane (TMS)	$(CH_3)_4Si$	88	74, 73, 55, 45, 43, 29
1,2-dimethoxy ethane	$(CH_3OCH_2)_2$	90	60, 58, 45 (100%), 31, 29
toluene	$C_6H_5CH_3$	92	91 (100%), 65, 51, 39, 28
chloroform	$CHCl_3$	118	120, 83, 81 (100%), 47, 35, 28
chlorodorm-d$_1$	$CDCl_3$	119	121, 84, 82 (100%), 48, 47, 35, 28
carbon tetrachloride	CCl_4	152 (not seen)	121, 119, 117 (100%), 84, 82, 58.5, 47, 35, 28
tetrachloroethene	$CCl_2=CCl_2$	164 (not seen)	168, 166, 165, 164, 131, 129, 128, 94, 82, 69, 59, 47, 31, 24

DETECTION OF LEAKS IN MASS SPECTROMETER SYSTEMS

The following four figures provide guidance for troubleshooting possible leaks in the vacuum systems of mass spectrometers, especially those operating in electron-impact mode. Leak testing is commonly done by playing a stream of a pure gas against a fitting, joint, or component that is suspected of being a leak source. If in fact the component is the source of a leak, one should be able to note the presence of the leak-detection fluid on the mass spectrum. Here we present the mass spectra of methane tetrafluoride, 1,1,1,2-tetrafluoroethane (R-134a), n-butane and acetone.[1,2] Methane tetrafluoride, 1,1,1,2-tetrafluoroethane, and n-butane are handled as gases, while acetone is handled as a liquid. Typically, n-butane is dispensed from a disposable lighter, and acetone is dispensed from a dropper. Care must be taken when using acetone or a butane lighter for leak checking because of the flammability of these fluids.

REFERENCES

1. Bruno, T.J., *CRC Handbook for the Analysis and Identification of Alternative Refrigerants*, CRC Press, Boca Raton, FL, 1994.
2. NIST, NIST Chemistry Web Book, NIST Standard Reference Database, No. 69, March 2003.

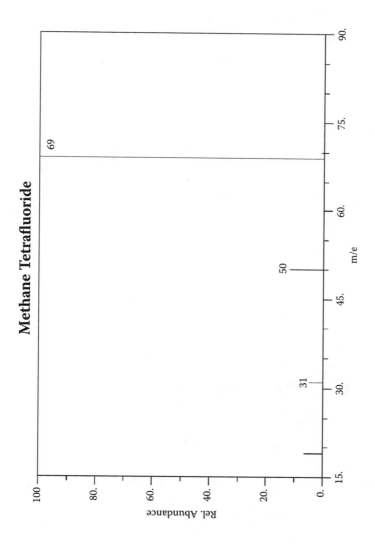

Methane Tetrafluoride

Rel. Abundance

m/e

Figure 4.1

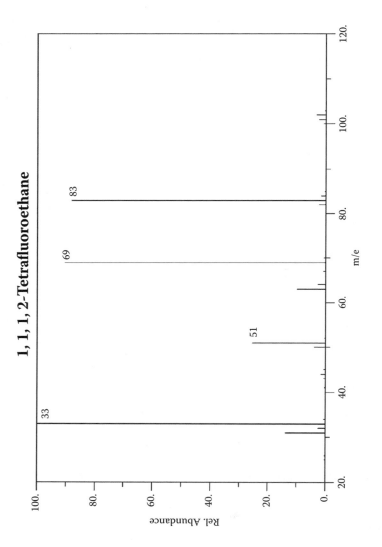

Figure 4.2

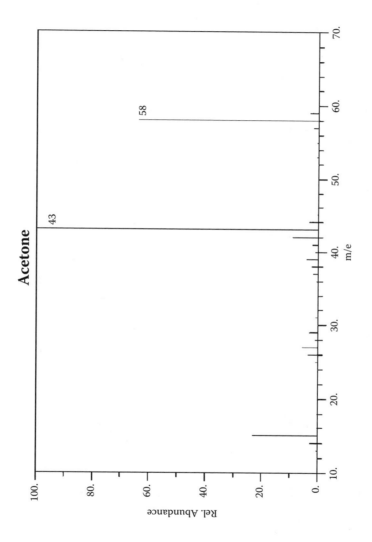

Figure 4.3

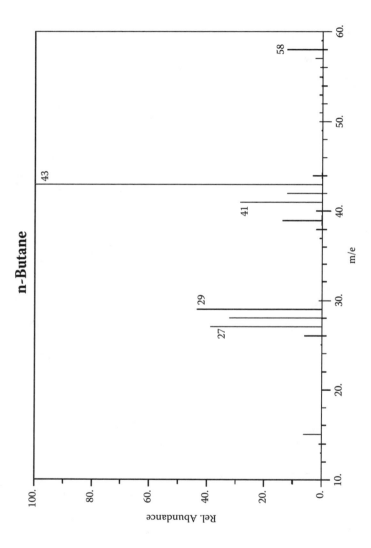

Figure 4.4

CHAPTER 5

Laboratory Safety

CONTENTS

MAJOR CHEMICAL INCOMPATIBILITIES

The following chemicals react, sometimes violently (indicated by italics), in certain chemical environments.[1-6] Incompatibilities may cause fires, explosions, or the release of toxic gases. Extreme care must be taken when working with these materials. This list is not inclusive, and the reader is urged to consult multiple sources for more specific information.

REFERENCES

1. Dean, J.A., Ed., *Lange's Handbook of Chemistry,* 15th ed., McGraw-Hill, New York, 1998.
2. Fieser, L.F. and Fieser, M., *Reagents for Organic Synthesis,* John Wiley and Sons, New York, 1967.
3. Gordon, A.J. and Ford, R.A., *The Chemist's Companion: A Handbook of Practical Data, Techniques, and References,* John Wiley and Sons, New York, 1972.
4. Shugar, G.J. and Dean, J.A., *The Chemist's Ready Reference Handbook,* McGraw-Hill, New York, 1990.
5. Svoronos, P., Sarlo, E., and Kulawiec, R., *Organic Chemistry Laboratory Manual,* 2nd ed., McGraw-Hill, New York, 1997.
6. Bruno, T.J. and Svoronos, P.D.N., *CRC Handbook of Basic Tables for Chemical Analysis,* 2nd ed., CRC Press, Boca Raton, FL, 2003.

Major Chemical Incompatibilities

Chemical	Incompatible Chemicals
acetic acid	Strong acids (chromic, nitric, perchloric), peroxides (see entry for carboxylic acids for more information)
acetylene	*Air*, copper, halogens (chlorine, bromine, iodine), alkali metals
alkali metals	*Acids, water, hydroxy compounds, polychlorinated hydrocarbons* (for example, CCl_4), halogens, carbon dioxide, oxidants, terminal alkynes
ammonia, anhydrous	*Halogens* (*bromine, chlorine, iodine*), *hydrofluoric acid, liquid oxygen*, calcium or sodium hypochlorite, heavy metals (silver, gold, mercury), nitric acid
ammonium nitrate	*Metal powders, chlorates, nitrites, sulfur, sugar*, flammable and combustible organics, acids, sawdust
anilines	*Concentrated acids* (*nitric, sulfuric, chromic*), *oxidizing agents* (*chromium ions*, peroxides, permanganate)
carbon, activated	Oxidizing agents, unsaturated oils
carboxylic acids	Metals (*alkalis*), organic bases, ammonia
chlorates	*Flammable and combustible organic compounds*, finely powdered metals, manganese dioxide, ammonium salts
chromic acid	Anilines, 1° or 2° alcohols, aldehydes
halogens (chlorine, bromine)	*Finely powdered metals, diethyl ether, hydrogen*, unsaturated organic compounds, carbide salts, acetylene, alkali metals
copper	Oxidizing agents
ether (diethyl)	*Peroxides* (especially after long exposure of ether to air)
fluorine	Reactive (as a strong oxidizing agent) to a certain degree with most compounds, but it can sometimes cause a violent reaction
hydrocarbons (saturated)	Halogens (especially *fluorine*) in the presence of ultraviolet light and peroxides
hydrocarbons (unsaturated)	Halogens, concentrated strong acids, peroxides, oxidizing agents
hydrofluoric acid	Ammonia, glass, organic bases
hydrogen peroxide	Metals, alcohols, potassium permanganate, flammable and combustible materials, unsaturated organics
iodine	*Acetaldehyde, antimony*, unsaturated hydrocarbons, ammonia, and some amines
mercury	Some metals, ammonia, terminal alkynes
nitric acid (concentrated)	*Anilines, flammable liquids, unsaturated organics*, lactic acid, coal, ammonia, amines, powdered metals, wood, alcohols, electron-rich aromatic rings (phenols)
perchloric acid	Some organics, acetic anhydride, metals, alcohols, wood and its derivatives, amines, inorganic bases
permanganates, general	Aldehydes, alcohols, unsaturated hydrocarbons
peroxides	*Flammable liquids, metals*, aldehydes, alcohols, impact, hydrocarbons (unsaturated)
picric acid	*Dryness and impact, alkali metals, oxidizing agents*, concentrated bases
potassium, metal	See entry for alkali metals
potassium permanganate	Hydrochloric acid, glycerol, hydrogen peroxide, sulfuric acid, wood, unsaturated hydrocarbons, alcohols
silver salts, organic	Dryness and prolonged air exposure
sodium, metal	See entry for alkali metals
sulfuric acid, concentrated	*Electron-rich aromatic rings* (*phenols, anilines*), *unsaturated hydrocarbons, potassium permanganate*, chlorates, perchlorates

ABBREVIATIONS USED IN THE ASSESSMENT AND PRESENTATION
OF LABORATORY HAZARDS

The following abbreviations are commonly encountered in presentations of laboratory and industrial hazards. The reader is urged to consult the reference below[1] for additional information.

REFERENCE

1. Furr, A.K., Ed., *CRC Handbook of Laboratory Safety,* 5th ed., CRC Press, Boca Raton, FL, 2000.

CC (closed cup): Method for measurement of the flash point. With this method, sample vapors are not allowed to escape, as they can with the open-cup (OC) method. Because of this, flash points measured with the CC method are usually a few degrees lower than those measured with the OC. The choice between CC and OC is dependent on the (usually ASTM) standard method chosen for the test.

COC (Cleveland open cup): See entry for OC (open cup).

IDLH (immediately dangerous to life and health): The maximum concentration of chemical contaminants, normally expressed as parts per million (ppm, mass/mass), from which one could escape within 30 minutes without a respirator, and without experiencing any escape-impairing (severe eye irritation) or irreversible health effects. Set by NIOSH. Note that this term is also used to describe electrical hazards.

LEL (lower explosion limit): The minimum concentration of a chemical in air at which detonation can occur.

LFL (lower flammability limit): The minimum concentration of a chemical in air at which flame propagation occurs.

MSDS (material safety data sheet): A (legal) document that must accompany any supplied chemical that provides information on chemical content, physical properties, hazards, and treatment of hazards. The MSDS should be considered only a minimal source of information, and cannot replace additional information available in other, more comprehensive sources.

NOEL (no observed effect level): The maximum dose of a chemical at which no signs of harm are observed. This term can also be used to describe electrical hazards.

OC (open cup, Cleveland open cup): This refers to the test method for determining the flash point of common compounds. It consists of a brass, aluminum, or stainless steel cup; a heater base to heat the cup; a thermometer in a fixture; and a test flame applicator. The flash point is the lowest temperature at which a material will form a flammable mixture with air above its surface. The lower the flash point, the easier the ignition.

PEL (permissible exposure level): An exposure limit that is published and enforced by OSHA as a legal standard. The PEL can be expressed as a time-weighted-average (TWA) exposure limit (for an 8-hour workday), a 15-minute short-term exposure limit (STEL), or a ceiling (C, or CEIL, or TLV-C).

RTECS (Registry of Toxic Effects of Chemical Substances): A database maintained by the National Institute of Occupational Safety and Health (NIOSH). The goal of the database is to include data on all known toxic substances, along with the concentration at which toxicity is known to occur. There are approximately 140,000 compounds listed.

STEL (short-term exposure level): An exposure limit for a short-term, 15-minute exposure that cannot be exceeded during the workday, enforced by OSHA as a legal standard. Short-term exposures below the STEL level generally will not cause irritation, chronic or irreversible tissue damage, or narcosis.

REL (recommended exposure level): Average concentration limit recommended for up to a 10-hour workday during a 40-hour workweek, by NIOSH.

TLV (threshold limit value): Guidelines suggested by the American Conference of Governmental Industrial Hygienists to assist industrial hygienists with limiting hazards of chemical exposures in the workplace.

TLV-C (threshold limit ceiling value): An exposure limit that should not be exceeded under any circumstances.

TWA (time-weighted average): Allowable concentration for a conventional 8-hour workday and a 40-hour workweek. It is the concentration to which it is believed that nearly all workers can be exposed without adverse health effects.

UEL (upper explosion limit): The maximum concentration of a chemical in air at which detonation can occur.

UFL (upper flammability limit): The maximum concentration of a chemical in air at which flame propagation can occur.

WEEL (workplace environmental exposure limit): Set by the American Industrial Hygiene Association (AIHA).

Some abbreviations that are sometimes used on material safety data sheets, and in other sources, are ambiguous. The most common meanings of some of these vague abbreviations are provided below, but the reader is cautioned that these are only suggestions:

Abbreviation	Meaning
EST	Established; estimated
MST	Mist (such as a fine spray)
N/A, NA	Not applicable
ND	None determined; not determined
NE	None established; not established
NEGL	Negligible
NF	None found; not found
N/K, NK	Not known
N/P, NP	Not provided
SKN	Skin
TS	Trade secret
UKN, UNK	Unknown

CHARACTERISTICS OF CHEMICAL-RESISTANT MATERIALS

The following table provides guidance in the selection of materials that provide some degree of chemical resistance for common laboratory tasks.[1]

REFERENCE

1. Furr, A.K., Ed., *CRC Handbook of Laboratory Safety*, 5th ed., CRC Press, Boca Raton, FL, 2000.

Characteristics of Chemical-Resistant Materials

Material	Abrasion Resistance	Cut Resistance	Flexibility	Heat Resistance	Ozone Resistance	Puncture Resistance	Tear Resistance	Relative Cost
butyl rubber	F	G	G	E	E	G	G	High
chlorinated polyethylene (CPE)	E	G	G	G	E	G	G	Low
natural rubber	E	E	E	F	P	E	E	Medium
nitrile-butadiene rubber (NBR)	E	E	E	G	F	E	G	Medium
neoprene	E	E	G	G	E	G	G	Medium
nitrile rubber (nitrile)	E	E	E	G	F	E	G	Medium
nitrile rubber + polyvinylchloride (nitrile + PVC)	G	G	G	F	E	G	G	Medium
polyethylene	F	F	G	F	F	P	F	Low
polyurethane	E	G	E	G	G	G	G	High
polyvinyl alcohol (PVA)	F	F	P	G	E	F	G	Very high
polyvinyl chloride (PVC)	G	P	F	P	E	G	G	Low
styrene-butadiene rubber (SBR)	E	G	G	G	F	F	F	Low
viton	G	G	G	G	E	G	G	Very high

Note: E = excellent, G = good, F = fair, P = poor.

SELECTION OF PROTECTIVE LABORATORY GARMENTS

The following table provides guidance in the selection of special protective garments that are used in the laboratory for specific tasks.[1]

REFERENCE

1. Hauck, P.G., *Personal Protective Equipment Guide*, Mount Sinai School of Medicine, New York, May 2002; available on-line at http://www.mssm.edu/biosafety/policies/pdfs/protective_equipment.pdf; accessed 5/19/05.

Selection of Protective Laboratory Garments

Material	Type of Garment	Common Use
Cotton and natural-fiber blends	Coveralls, lab coats, sleeve protectors, aprons	For dry dusts, particulates, and aerosols
Tyvek	Coveralls, lab coats, sleeve protectors, aprons, hoods	For dry dusts and aerosols
Saranax/Tyvek SL	Coveralls, lab coats, sleeve protectors, aprons, hoods, level-B suits	For aerosols, liquids, and solvents
Polyethylene	Barrier gowns, aprons	For body fluids
Polypropylene	Clean-room suits, coveralls, lab coats	For dry dusts and nontoxic particulates
Polyethylene/Tyvek QC	Coveralls, aprons, lab coats, shoe covers	For moisture and solvents
Polypropylene	Coveralls, lab coats, shoe covers, caps, clean-room suits	For nontoxic particulates and dry dusts
Tychem BR, Tychem TK	Full level-A and level-B suits	For highly toxic particulates and dry dusts
CPF	Full level-A and level-B suits, splash suits	For highly toxic chemicals, gases, aerosols
Polyvinyl chloride	Full level-A suits	For highly toxic chemicals, gases, and aerosols

PROTECTIVE CLOTHING LEVELS

In the United States, OSHA defines various levels of protective clothing and sets parameters that govern their use with chemical spills and in environments where chemical exposure is a possibility. A summary of the definitions is provided below.[1]

REFERENCE

1. OSHA, OSHA Technical Manual, Section VIII, Chapter 1, Chemical Protective Clothing, 2003; available on-line at http://www.osha.gov/dts/osta/otm/otm_viii/otm_viii_1.html; accessed 5/19/05.

Protective Clothing Levels

Description	Protection Provided	When Used	Limitations

Level A

Vapor-protective suit (meets NFPA 1991); pressure-demand, full-face piece Self-contained breathing apparatus (SCBA); inner chemical-resistant gloves and chemical-resistant safety boots; two-way radio communications	Highest available level of respiratory, skin, and eye protection from solid, liquid, and gaseous chemicals	The chemical(s) have been identified and have a high level of hazards to respiratory system, skin, and eyes; substances are present with known or suspected skin toxicity or carcinogenicity; operations must be conducted in confined or poorly ventilated areas	Protective clothing must resist permeation by the chemicals or mixtures present

Level B

Liquid splash-protective suit (meets NFPA 1992); pressure-demand, full-face piece SCBA; inner chemical-resistant gloves and chemical-resistant safety boots; two-way radio communications	Provides same level of respiratory protection as Level A, but somewhat less skin protection; liquid splash protection is provided, but not protection against chemical vapors or gases	The chemical(s) have been identified but do not require a high level of skin protection; the primary hazards associated with site entry are from liquid and not vapor contact	Protective clothing items must resist penetration by the chemicals or mixtures present

Level C

Support-function protective garment (meets NFPA 1993); full-face piece, air-purifying, canister-equipped respirator; chemical-resistant gloves and safety boots; two-way communications system	The same level of skin protection as Level B, but a lower level of respiratory protection; liquid splash protection but no protection to chemical vapors or gases	Contact with site chemical(s) will not affect the skin; air contaminants have been identified and concentrations measured; a canister is available that can remove the contaminant; the site and its hazards have been completely characterized	Protective clothing items must resist penetration by the chemicals or mixtures present; chemical airborne concentration must be less than IDLH levels; the atmosphere must contain at least 19.5% oxygen; not acceptable for chemical emergency response

Level D

Coveralls, safety boots/shoes, safety glasses or chemical-splash goggles	No respiratory protection; minimal skin protection	The atmosphere contains no known hazard; work functions preclude splashes, immersion, potential for inhalation, or direct contact with hazard chemicals	The atmosphere must contain at least 19.5% oxygen; not acceptable for chemical emergency response

Optional items can be added to each level of protective clothing. Options include items from higher levels of protection as well as hard hats, hearing protection, outer gloves, a cooling system, etc.

SELECTION OF LABORATORY GLOVES

The following table provides guidance in the selection of protective gloves for laboratory use.[1-5] If protection from more than one class of chemical is required, double gloving should be considered.

REFERENCES

1. Garrod, A.N., Martinez, M., and Pearson, J., *Ann. Occup. Hyg.*, 43, 543–55, 1999.
2. Garrod, A.N., Phillips, A.M., and Pemberton, J.A., *Ann. Occup. Hyg.*, 45, 55–60, 2001.
3. Mockelsen, R.L. and Hall, R.C., *Am. Ind. Hyg. Assoc. J.*, 48, 941–947, 1987.
4. OSHA, *Federal Register,* 59 (66), 16334-16364, 29 CFR 1910, 1994.
5. Bruno, T.J. and Svoronos, P.D.N., *CRC Handbook of Basic Tables for Chemical Analysis,* 2nd ed., CRC Press, Boca Raton, FL, 2003.

Selection of Laboratory Gloves

Glove Material	Resistant To
Viton	PCBs, chlorinated solvents, aromatic solvents
Viton/butyl	Acetone, toluene, aromatics, aliphatic hydrocarbons, chlorinated solvents, ketones, amines, and aldehydes
SilverShield and 4H (PE/EVAL)	Morpholine, vinyl chloride, acetone, ethyl ether, many toxic solvents, and caustics
Barrier	Wide range of chlorinated solvents and aromatic acids
PVA	Ketones, aromatics, chlorinated solvents, xylene, methyl isobutyl ketone, trichloroethylene; *do not use with water/aqueous solutions*
Butyl	Aldehydes, ketones, esters, alcohols, most inorganic acids, caustics, dioxane
Neoprene	Oils, grease, petroleum-based solvents, detergents, acids, caustics, alcohols, miscellaneous organic solvents
PVC	Acids, caustics, organic solvents, grease, oil
Nitrile	Oils, fats, acids, caustics, alcohols
Latex	Body fluids, blood, acids, alcohols, alkalis
Vinyl	Body fluids, blood, acids, alcohols, alkalis
Rubber	Organic acids, some mineral acids, caustics, alcohols; not recommended for aromatic solvents, chlorinated solvents

SELECTION OF RESPIRATOR CARTRIDGES AND FILTERS

Respirators are sometimes desirable or required when performing certain tasks in the chemical analysis laboratory. There is a standardized color-code system used by all manufacturers for the specification and selection of the cartridges and filters that are used with respirators. The following table[1] provides guidance in the selection of the proper cartridge using the color code.

REFERENCE

1. Bruno, T.J. and Svoronos, P.D.N., *CRC Handbook of Basic Tables for Chemical Analysis,* 2nd ed., CRC Press, Boca Raton, FL, 2003.

Selection of Respirator Cartridges and Filters

Color Code	Application
Gray	Organic vapors, ammonia, methylamine, chlorine, hydrogen chloride, and sulfur dioxide or hydrogen sulfide (for escape only), or hydrogen fluoride or formaldehyde
Black	Organic vapors, not to exceed regulatory standards
Yellow	Organic vapors, chlorine, chlorine dioxide, hydrogen chloride, hydrogen fluoride, sulfur dioxide, or hydrogen sulfide (for escape only)
White	Chlorine, hydrogen chloride, hydrogen fluoride, sulfur dioxide, or hydrogen sulfide (for escape only)
Green	Ammonia and methylamine
Orange	Mercury or chlorine
Purple	Solid and liquid aerosols and mists
Purple + gray	Organic vapors, ammonia, methylamine, chlorine, hydrogen chloride, and sulfur dioxide or hydrogen sulfide (for escape only) or hydrogen fluoride or formaldehyde; solid and liquid aerosols and mists
Purple + black	Organic vapors and solid and liquid aerosols and mists
Purple + yellow	Organic vapors, chlorine, chlorine dioxide, hydrogen chloride, hydrogen fluoride, sulfur dioxide, or hydrogen sulfide (for escape only); solid and liquid aerosols and mists
Purple + white	Chlorine, hydrogen chloride, hydrogen fluoride, sulfur dioxide, or hydrogen sulfide (for escape only); solid and liquid aerosols and mists
Purple + green	Ammonia, methylamine, and solid and liquid aerosols and mists

In addition to the cartridges specified in the table, particulate filters are available that can be used alone or in combination.

STANDARD CGA FITTINGS FOR COMPRESSED-GAS CYLINDERS

The following table presents a partial list of gases and the CGA fittings that are required to use those gases when they are stored in, and dispensed from, compressed-gas cylinders.[1]

REFERENCE

1. CGA Pamphlet V-1-99, American Canadian and Compressed Gas Association Standard for Compressed Gas Cylinder Valve Outlet and Inlet Connections, ANSI B57.1, CSA B96, 1999, revision sheet 2004.

Standard CGA Fittings for Compressed-Gas Cylinders

Gas	Fitting
acetylene	510
air	346
carbon dioxide	320
carbon monoxide	350
chlorine	660
ethane	350
ethylene	350
ethylene oxide	510
helium	580
hydrogen	350
hydrogen chloride	330
methane	350
neon	580
nitrogen	580
nitrous oxide	326
oxygen	540
sulfur dioxide	660
sulfur hexafluoride	590
xenon	580

Source: Reproduced from the CGA Pamphlet V-1-87, American Canadian and Compressed Gas Association Standard for Compressed Gas Cylinder Valve Outlet and Inlet Connections, ANSI B57.1, CSA B96, by permission of the Compressed Gas Association.

The following graphic shows the geometry and dimensions of common CGA fittings for compressed-gas cylinders.

Connection 110 - Lecture bottle outlet
for corrosive gases - 5/16" - 32 RH INT.,
with gasket

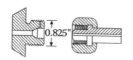

Connection 326 - 0.825" - 14 RH EXT.

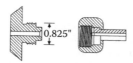

Connection 170 - Lecture bottle outlet for non-corrosive gases
9/16" - 18 RH EXT. and 5/16" - 32 RH INT., with gasket

Connection 350 - 0.825" - 14 LH EXT.

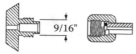

Connection 320 - 0.825" - 14 RG EXT., with gasket

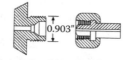

Connection 540 - 0.903" - 14 RH EXT.

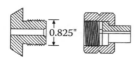

Connection 330 - 0.825" - 14 LH EXT., with gasket

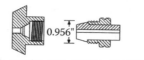

Connection 590 - 0.956" - 14 LH INT.

Connection 510 - 0.885" - 14 LH INT.

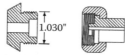

Connection 660 - 1.030" - 14 RH EXT., with gasket

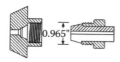

Connection 580 - 0.965" - 14 RH INT.

Figure 5.1

GAS CYLINDER STAMPED MARKINGS

The graphic below describes the permanent, stamped markings that are used on high-pressure gas cylinders commonly found in analytical laboratories. Note that individual jurisdictions and institutions have requirements for marking the cylinder contents as well. These requirements are in addition to the stamped markings, which pertain to the cylinder itself rather than to the fill contents.

There are four fields of markings on cylinders that are used in the United States, labeled 1 through 4 on the figure.

Field 1 — cylinder specifications: DOT stands for the U.S. Department of Transportation, the agency that regulates the transport and specification of gas cylinders in the United States. The next entry (e.g., 3AA in the figure) is the specification for the type and material of the cylinder. The most common cylinders are 3A, 3AA, 3AX, 3AAX, 3T, and 3AL. All but the last refer to steel cylinders, while 3AL refers to aluminum. The individual specifications differ mainly in chemical composition of the steel and the gases that are approved for containment and transport. The 3T deals with large bundles of tube trailer cylinders. The next entry in this field (e.g., 2015 in the figure) is the service pressure, in psig.

Field 2 — serial number: This is a unique number assigned by the manufacturer.

Field 3 — identifying symbol: The manufacturer-identifying symbol historically can be a series of letters or a unique graphical symbol. In recent years, the DOT has standardized this identification with the "M" number, for example, M1004. This is a number issued by DOT that identifies the cylinder manufacturer.

Field 4 — manufacturing data: The data of manufacture is provided as a month and year. With this date is the inspector's official mark, for example, H. In recent years, this letter has been replaced with an IA number, for example IA02, pertaining to an independent agency that is approved by DOT as an inspector.

If "+" is present, the cylinder qualifies for an overfill of 10% in service pressure.

If "♠" is present, the cylinder qualifies for a 10-year rather than a 5-year retest interval.

Also stamped on the cylinder will be the retest dates. A cylinder must have a current (that is, within 5 or 10 years) test stamp. The owner of the cylinder may also be stamped on the collar of the cylinder.

REFERENCE

1. Hazardous Materials: Requirements for Maintenance, Requalification, Repair and Use of DOT Specification Cylinders, 49 CFR Parts 107, 171, 172, 173, 177, 178, 179, and 180, Docket No. RSPA-01-10373 (HM-220D), RIN 2137-AD58, August 8, 2002. http://www.gpoaccess.gov/cfr/index.html.

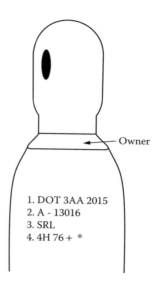

1. DOT 3AA 2015
2. A - 13016
3. SRL
4. 4H 76 + *

Figure 5.2

CHAPTER 6

Unit Conversions and Physical Constants

CONTENTS

UNIT CONVERSIONS

The International System of Units (SI) is described in detail in NIST Special Publication 811[1] and in lists of physical constants and conversion factors.[2-6] Selected unit conversions[1-6] are given in the following tables. The conversions are presented in matrix format when all of the units are of a convenient order of magnitude. When some of the unit conversions are of little value (such as the conversion between metric tons and grains), tabular form is followed, with the less useful units omitted.

REFERENCES

1. Taylor, B.N., *Guide for the Use of the International System of Units,* Special Publication SP-811, National Institute of Standards and Technology, Washington, DC, 1995.
2. Chiu, Y., *A Dictionary for Unit Conversion,* School of Engineering and Applied Science, George Washington University, Washington, DC, 1975.
3. Lide, D.R., Ed., *CRC Handbook for Chemistry and Physics,* 83rd ed., CRC Press, Boca Raton, FL, 2003.
4. Bruno, T.J. and Svoronos, P.D.N., *CRC Handbook of Basic Tables for Chemical Analysis,* 2nd ed., CRC Press, Boca Raton, FL, 2003.
5. A Dictionary of Units. http://www.ex.ac.uk/cimt/dictunit/dictunit.htm.
6. Kimball's Biology Pages, 2004; available on-line at http://biology_pages.info. http://users.rcn.com/jkimball.ma.ultranet/BiologyPages/U/Unit.html.

Area

Multiply	By	To Obtain
Square millimeters	0.00155	square inches (U.S.)
	1×10^{-6}	square meters
	0.01	square centimeters
	1.2732	circular millimeters
Square centimeters	1.196×10^{-4}	square yards
	0.00108	square feet
	0.15500	square inches
	1×10^{-4}	square meters
	100	square millimeters
Square kilometers	0.38610	square miles (U.S.)
	1.1960×10^{6}	square yards
	1.0764×10^{7}	square feet
	1×10^{6}	square meters
	247.10	acres (U.S.)
Square inches (U.S.)	0.00694	square feet
	0.00077	square yards
	6.4516×10^{-4}	square meters
	6.4516	square centimeters
	645.15	square millimeters
Square feet (U.S.)	3.5870×10^{-8}	square miles
	0.11111	square yards
	144	square inches
	0.09290	square meters
	929.03	square centimeters
	2.2957×10^{-5}	acres
Square miles	640	acres
	3.0967×10^{6}	square yards
	2.7878×10^{7}	square feet
	2.5900	square kilometers

Density

kg/m^3	g/cm^3	lb/ft^3
16.018	0.016018	1
1	0.001	0.062428
1000	1	62.428
2015.9	2.0159	125.85

Enthalpy, Heat of Vaporization, Heat of Conversion, Specific Energies

kJ/kg (J/g)	cal/g	Btu/lb
2.3244	0.55556	1
1	0.23901	0.43022
4.1840	1	1.8

Length

Multiply	By	To Obtain
angstroms	1×10^{-10}	meters
	3.9370×10^{-9}	inches (U.S.)
	1×10^{-4}	micrometers
	1×10^{-8}	centimeters
	0.1	nanometers
nanometers (nm)	1×10^{-9}	meters
	1×10^{-7}	centimeters
	10	angstroms
micrometers (μm)	3.9370×10^{-5}	inches (U.S.)
	1×10^{-6}	meters
	1×10^{-4}	centimeters
	1×10^{4}	angstroms
millimeters	0.03937	inches (U.S.)
	1×10^{3}	micrometers
centimeters	0.39370	inches (U.S.)
	1×10^{4}	micrometers (:m)
	1×10^{7}	nanometers
	1×10^{8}	angstroms
meters	6.2137×10^{-4}	miles (statute)
	1.0936	yards (U.S.)
	39.370	inches (U.S.)
	1×10^{9}	millimicrons
	1×10^{10}	angstroms
kilometers	0.53961	miles (nautical)
	0.62137	miles (statute)
	1093.6	yards
	3280.8	feet
inches (U.S.)	0.02778	yards
	2.5400	centimeters
	2.5400×10^{8}	angstroms
feet (U.S.)	0.30480	meters
	30.480	centimeters
yards (U.S.)	5.6818×10^{-4}	miles
	0.91440	meters
	91.440	centimeters
miles (nautical)	1.1516	statute miles
	2026.8	yards
	1.8533	kilometers
miles (U.S. statute)	320	rods
	0.86836	nautical miles
	1.6094	kilometers
	1609.4	meters

Parts Per Million

ppm	vs.	Percent
1 ppm	=	0.0001%
10 ppm	=	0.001%
100 ppm	=	0.01%
1000 ppm	=	0.1%
10,000 ppm	=	1.0%
100,000 ppm	=	10.0%
1,000,000 ppm	=	100.0%

Parts Per Billion

ppb	vs.	Percent
10	=	0.000 001%
100	=	0.000 01%
1000	=	0.0001%
10,000	=	0.001%
100,000	=	0.01%
1,000,000	=	0.1%

Parts Per Trillion

ppt	vs.	Percent
100	=	1×10^{-8}%
10,000	=	0.000001%
1,000,000	=	0.0001%
100,000,000	=	0.01%

The following table provides guidance in the use of base-ten concentration units (presented in the three preceding tables), since there are differences in usage.

Concentration Units Nomenclature

Number	Number of Zeros	Name (Scientific Community)	Name (U.K., France, Germany)
1000	3	thousand	thousand
1,000,000	6	million	million
1,000,000,000	9	billion	milliard, or thousand million
1,000,000,000,000	12	trillion	billion
1,000,000,000,000,000	15	quadrillion	thousand billion

Pressure

MPa	atm	torr (mm Hg)	bar	lbs/in.² (psi)
6.8948×10^{-3}	0.068046	51.715	6.8948×10^{-2}	1
1	9.8692	7500.6	10.0	145.04
0.101325	1	760.0	1.01325	14.696
1.3332×10^{-4}	1.3158×10^{-3}	1	1.332×10^{-3}	0.019337
0.1	0.98692	750.06	1	14.504

Specific Heat, Entropy

kJ/(kg · K), J/(g · K)	Btu/(°R · lb)
4.184	1
1	0.23901

Specific Volume

m³/kg (L/g)	cm³/g	ft³/lb
0.062428	62.428	1
1	1000	16.018
0.001	1	0.016018

Surface Tension

N/m	dyne/cm	lb/in.
175.13	175.13×10^3	1
1	1000	5.7102×10^{-6}
0.001	1	5.7102×10^{-3}

Temperature

T(rankine)	=	1.8T(kelvins)
T(celsius)	=	T(kelvins) −273.15
T(fahrenheit)	=	T(rankine) −459.67
T(fahrenheit)	=	1.8T(celsius) +32

Thermal Conductivity

mW/(cm · K)	J/(s·cm · K)	cal/(s·cm · K)	Btu/(ft·hr · °R)
17.296	0.017296	0.0041338	1
1	0.001	2.3901×10^{-4}	0.057816
1000	1	0.23901	57.816
4184	4.184	1	241.90

Velocity

Multiply	By	To Obtain
feet per minute	0.01136	miles per hour
	0.01829	kilometers per hour
	0.5080	centimeters per second
	0.01667	feet per second
feet per second	0.6818	miles per hour
	1.097	kilometers per hour
	30.48	centimeters per second
	0.3048	meters per second
	0.5921	knots
knots (Br)	1.0	nautical miles per hour
	1.6889	feet per second
	1.1515	miles per hour
	1.8532	kilometers per hour
	0.5148	meters per second
meters per second	3.281	feet per second
	2.237	miles per hour
	3.600	kilometers per hour
miles per hour	1.467	feet per second
	0.4470	meters per second
	1.609	kilometers per hour
	0.8684	knots

Velocity of Sound

m/s	ft/s
0.3048	1
1	3.2808

Viscosity

kg/(m · s) (N·s/m², Pa · s)	cP (10⁻²g/(cm · s))	lb·s/ft² (slug/(ft · s))	lb/(ft · s)
1.48816	1488.16	0.31081	1
1	1000	0.020885	0.67197
0.001	1	2.0885×10^{-5}	6.7197×10^{-4}
47.881	4.7881×10^{-4}	1	32.175

Volume

Multiply	By	To Obtain
barrels (pet)	42	gallons (U.S.)
	34.97	gallons (Br.)
cubic centimeters	10^{-3}	liters
	0.0610	cubic inches
cubic feet	28,317	cubic centimeters
	1,728	cubic inches
	0.03704	cubic yards
	7.481	gallons (U.S., liq.)
	28.317	liters
cubic inches	6.387	cubic centimeters
	0.016387	liters
	4.329×10^{-3}	gallons (U.S. liq.)
	0.01732	quarts (U.S. liq.)
gallons, imperial	277.4	cubic inches
	1.201	U.S. gallons
	4.546	liters
gallons, U.S. (liquid)	231	cubic inches
	0.1337	cubic feet
	3.785	liters
	0.8327	imperial gallons
	128	fluid ounces (U.S.)
ounces, fluid	29.57	cubic centimeters
	1.805	cubic inches
liters	0.2642	gallons
	0.0353	cubic feet
	1.0567	quarts (U.S. liq.)
	61.025	cubic inches
quarts, U.S. (liquid)	0.0334	cubic feet
	57.749	cubic inches
	0.9463	liters

Mass (Weight)

Multiply	By	To Obtain
milligrams	2.2046×10^{-6}	pounds (avoirdupois)
	3.5274×10^{-5}	ounces (avoirdupois)
	0.01543	grains
	1×10^{-6}	kilograms
micrograms	1×10^{-6}	grams
grams	0.00220	pounds (avoirdupois)
	0.03527	ounces (avoirdupois)
	15.432	grains
	1×10^{6}	micrograms
kilograms	0.00110	tons (short)
	2.2046	pounds (avoirdupois)
	35.274	ounces (avoirdupois)
	1.5432×10^{4}	grains
grains	1.4286×10^{-4}	pounds (avoirdupois)
	0.00229	ounces (avoirdupois)
	0.06480	grams
	64.799	milligrams
ounces (avoirdupois)	3.1250×10^{-5}	tons (short)
	0.06250	pounds (avoirdupois)
	437.50	grains
	28.350	grams
pounds (avoirdupois)	5×10^{-4}	tons (short)
	16	ounces (avoirdupois)
	7000	grains
	0.45359	kilograms
	453.59	grams
tons (short, U.S.)	2000	pounds (avoirdupois)
	3.200×10^{4}	ounces (avoirdupois)
	907.19	kilograms
tons (long)	2240	pounds (avoirdupois)
	1016	kilograms
tons (metric)	1000	kilograms
	2205	pounds (avoirdupois)
	1.102	tons (short)

Prefixes for SI Units

Fraction	Prefix	Symbol
10^{-1}	deci	d
10^{-2}	centi	c
10^{-3}	milli	m
10^{-6}	micro	μ
10^{-9}	nano	n
10^{-12}	pico	p
10^{-15}	femto	f
10^{-18}	atto	a

Multiple	Prefix	Symbol
10	deka	da
10^2	hecto	h
10^3	kilo	k
10^6	mega	M
10^9	giga	G
10^{12}	tera	T
10^{15}	peta	P
10^{18}	exa	E

RECOMMENDED VALUES OF SELECTED PHYSICAL CONSTANTS

The following table provides some commonly used physical constants that are of value in thermodynamic and spectroscopic calculations.[1,2]

REFERENCES

1. Lide, D.R., Ed., *CRC Handbook for Chemistry and Physics,* 83rd ed., CRC Press, Boca Raton, FL, 2003.
2. NIST, NIST Reference on Constants, Units and Uncertainty, 2004; available on-line at http://physics .nist.gov/cuu/; accessed 5/20/05.

Values of Selected Physical Constants

Physical Constant	Symbol	Value
Avogadro constant	N_A	$6.02214199 \times 10^{23}$ mol^{-1}
Boltzmann constant	k	$1.3806503 \times 10^{-23}$ J K^{-1}
Charge-to-mass ratio	e/m	$-1.758820174 \times 10^{11}$ C kg^{-1}
Elementary charge	e	1.60218×10^{-19} C
Faraday constant	F	96485.3415 C mol^{-1}
Molar gas constant	R	8.314472 J mol^{-1} K^{-1}
"Ice point" temperature	T_{ice}	273.150 K (exactly)
Molar volume of ideal gas (STP)	V_m	2.24138×10^{-2} m^3 · mol^{-1}
Permittivity of vacuum	ε_o	8.854188×10^{-12} kg^{-1} m^{-3} · s^4 · A^2 (F · m^{-1})
Planck constant	h	$6.626068\ 76 \times 10^{-34}$ J · s
Standard atmosphere pressure	p	101325 N · m^{-2} (exactly)
Atomic mass constant	m_u	$1.660538\ 73 \times 10^{-27}$ kg
Speed of light in vacuum	c	299,792,458 m · s^{-1} (exactly)

Subject Index

A

O

P

Q

R

Rare earth filters 15
Recommended exposure level, 190
Recommended values of selected physical constants, 211
Reference compounds, electron impact mass spectrometry, 175
Reference standards, NMR, 87
Registry of Toxic Effects of Chemical Substances, 190
REL, 190
Respiratory cartridges, 197
RTECS, 190
Rules for determination of molecular formula, 169

S

Salts, NMR absorption, ^{1}H, 103
Selenium compounds, inorganic, infrared absorption, 38–39
Short term exposure level, 190
Silicon
     compounds, inorganic, infrared absorption, 38–39
     isotope abundance, 168
^{29}Si NMR absorptions, 157
^{119}Si NMR absorptions, 162
SI units, prefixes for, 210–211
Solubility parameter, 40
Solvatochromic parameters, 40
Solvent, for ultraviolet spectrophotometry, 11
Specific energies, unit conversion, 205
Specific heat, units conversion, 207
Specific volume, units conversion, 208
Speed of light in vacuum, recommended, 211
Spin-spin coupling to ^{15}N, 136
sp^3 nitrogen, NMR absorptions, ^{15}N, 134
Square centimeters, unit conversion, 205
Square feet, unit conversion, 205
Square inches, unit conversion, 205
Square kilometers, unit conversion, 205
Square miles, unit conversion, 205
Square millimeters, unit conversion, 205
Square yards, unit conversion, 205
S-substituted compounds, NMR absorption, ^{1}H, 98, 101, 104, 107
Standard atmosphere pressure, recommended, 211
STEL, 190
Strong acids, incompatibilities with, 189
Sulfides
     fragmentation in mass spectrometry, 178
     NMR absorption, ^{1}H, 91, 94, 98, 101, 104, 107
Sulfonamides, fragmentation in mass spectrometry, 179
Sulfones
     fragmentation in mass spectrometry, 179
     NMR absorption, ^{1}H, 91, 94
     UV–Vis absorption, 3
Sulfonic acid and derivatives, NMR absorption, ^{1}H, 91, 94, 98, 101, 104, 107
Sulfonium salts, NMR absorption, ^{1}H, 91, 94

Sulfoxides
     fragmentation in mass spectrometry, 179
     NMR absorption, ^{1}H, 91, 94
     UV–Vis absorption, 5
Sulfur
     compounds
         inorganic, infrared absorption, 38–39
         NMR absorption, ^{13}C, 129, 131
     heterocycles, NMR absorption, ^{1}H, 110, 112, 114, 116
     isotope abundance, 168
Surface tension, units conversion, 208

T

Temperature, units conversion, 208
Tetracoordinated phosphorus compounds, NMR absorptions, ^{31}P, 152, 153
Thermal conductivity, units conversion, 208
Thiocyanates, NMR absorption, ^{1}H, 91, 94, 98, 101, 103, 104, 107
Thioethers
     fragmentation in mass spectrometry, 178
     UV–Vis absorption, 7
Thioketone, UV–Vis absorption, 3
Thiols
     fragmentation in mass spectrometry, 179
     NMR absorption, ^{1}H, 91, 94, 98, 101, 104, 107
     UV–Vis absorption, 7
Threshold limit ceiling value, 191
Threshold limit value, 190
Time-weighted average, 191
TLV, 190
TLV-C, 191
Tons
     long, units conversion, 210
     metric, units conversion, 210
     short, units conversion, 210
Transmittance, absorbance conversion, 13
Tricoordinated phosphorus compounds, NMR absorptions, ^{31}P, 150, 151
Tungsten compounds, inorganic, infrared absorption, 38–39
TWA, 191

U

UEL, 191
UFL, upper flammability limit, 191
Ultraviolet spectrophotometry, solvents for, 11
Ultraviolet-visible region, 2
Ultraviolet-visible spectrophotometers, calibration of, 15
Ultraviolet–visible spectrophotometry, 1–16
     calibration of ultraviolet–visible spectrophotometers, 15–16
     correlation charts, 2–7
     solvents, 11–12
     transmittance–absorbance conversion, 13–14
     Woodward's rule for bathochromic shifts, 8–10
Unit conversions and physical constants, 203–211

Chemical Compound Index